Provisional atlas of the aculeate Hymenoptera of Britain and Ireland Part 9

Graham A. Collins (Editor)
Bees, Wasps and Ants Recording Society
Helen E. Roy (Editor)
Biological Records Centre

Citation information:

Collins, G.A. and Roy, H.E., eds. 2016. *Provisional atlas of the aculeate Hymenoptera of Britain and Ireland. Part 9*. Biological Records Centre, Wallingford.

CONTENTS

	Page
Acknowledgements	2
Introduction	3
Systematic list of species mapped	5
Distribution maps and species profiles	7
Dryinidae	8
Formicidae	14
Pompilidae	30
Crabronidae	38
Sphecidae	48
Andrenidae	50
Apidae	82
Halictidae	94
Megachilidae	112
References	121
List of plant names	127
Cumulative index to species in Provisional Atlas parts 1 to 9	129

ACKNOWLEDGEMENTS

This ninth part of the Bees, Wasps and Ants Provisional Atlas is the result of much hard work by many people.

The species for this Atlas were carefully selected by Target Species Coordinator, Stuart Roberts, together with the following specialist group compilers: Geoff Allen (Halictidae), Michael Archer (Dryinidae), John Burn (Dryinidae), Graham Collins (Pompilidae), Mike Edwards (Andrenidae, Apidae), George Else (Andrenidae, Apidae), Mike Fox (Formicidae), Andy Jukes (Megachilidae) and Adrian Knowles (Crabronidae, Sphecidae).

The species profiles written by the compilers have been edited by Graham Collins (BWARS) and Helen Roy (BRC).

The large number of entomologists and naturalists who submitted records for the target species are thanked for their invaluable contributions. Last, but by no means least, Mike Edwards and Stuart Roberts have undertaken the huge task of checking all the data and producing the maps. These were drawn on computer using the DMAP mapping package developed by Alan J Morton.

Our thanks go to all these contributors.

We are especially indebted to The Trustees of the Natural History Museum, London, who have made their collections of aculeates available for study by our recorders.

INTRODUCTION

Part 9 of the *Provisional Atlas of aculeate Hymenoptera* marks the beginning of the end for the atlas series. It was always intended to cover the British species in ten parts and whilst this part and the next cover some species that are widespread and common they also include some that are restricted to the Channel Islands, as well as some that have only been recorded from Britain on a few occasions.

With the addition of this part to the series we will have covered just over 80% of the British species of aculeate (see Table 1). One further part is envisaged, which will cover some, but perhaps not all, of the remainder. Although three species of dryinid are included in this part, and a couple of bethylids together with our single embolemid have been covered in earlier parts, it is unlikely that all remaining species can be covered. The fact is that they remain a highly specialised group, with very few recorders collecting them and even fewer able to identify them. There are several options being considered to deal with them.

Whatever their fate, one thing is certain – we will not be publishing a single volume updating all the maps.

With rapidly changing distributions and growing interest in studying aculeates leading to improved knowledge of their biology it is clear that the internet presents the best medium to disseminate this knowledge. As the atlas series draws to a close we shall provide updated species accounts on the society's website (www.bwars.com). Together with distribution maps, which can be viewed there already, this will provide the best means of keeping the species accounts current.

Eight parts of the *Provisional Atlas* have been published so far:

Part 1 with 55 species (Edwards, 1997)

Part 2 with 55 species (Edwards, 1998)

Part 3 with 60 species (Edwards & Telfer, 2001)

Part 4 with 55 species (Edwards & Telfer, 2002)

Part 5 with 60 species (Edwards & Broad, 2005)

Part 6 with 59 species (Edwards & Broad, 2006)

Part 7 with 58 species (Edwards & Roy, 2009)

Part 8 with 59 species (Collins & Roy, 2012)

The atlas coverage by family is shown in Table 1.

Table 1. Breakdown of the British and Irish aculeate fauna by family, showing the coverage of the first nine parts of the atlas. Figures based on the BWARS checklist published in the Members' Handbook (Else in Archer, 2005).

Family	Total no. of species	No. spp mapped	No. left to map
Dryinidae	34	6	28
Embolemidae	1	1	0
Bethylidae	22	2	20
Chrysididae	33	33	0
Tiphiidae	3	3	0
Mutillidae	3	3	0
Sapygidae	2	2	0
Scoliidae	1	0	1
Formicidae	54	43	10
Pompilidae	45	41	4
Vespidae	34	31	3
Sphecidae	7	5	2
Crabronidae	123	114	9
Apidae (s.l.)	268	233	35
Total	**630**	**517**	**112**

In this ninth part of the *Provisional Atlas*, 56 species are added, taking the total number mapped to 517. All the species covered here have previously been published in our newsletter, allowing feedback from our members.

SYSTEMATIC LIST OF SPECIES MAPPED

Previous Atlas parts have used the classification as published in the BWARS Members' Handbook (*Else* in Archer, 2005). Part 9 (and the society's website) have now adopted a new classification based on the checklist of Else, Bolton & Broad (2016). The principal difference is that the subfamilies of bees have been raised to family level. Changes in individual names are given in the species accounts. Map numbers are given for each species.

HYMENOPTERA ACULEATA
CHRYSIDOIDEA
DRYINIDAE
Anteoninae
    462   *Anteon ephippiger*
    463   *Anteon gaullei*
    464   *Anteon infectum*
VESPOIDEA
FORMICIDAE
Formicinae
    465   *Lasius alienus*
    466   *Lasius mixtus*
    467   *Lasius niger*
    468   *Lasius platythorax*
    469   *Lasius psammophilus*
    470   *Lasius sabularum*
    471   *Lasius umbratus*
Myrmicinae
    472   *Myrmica rubra*
POMPILIDAE
Pepsinae
    473   *Priocnemis confusor*
    474   *Priocnemis hyalinata*
    475   *Priocnemis parvula*
    476   *Priocnemis pusilla*
APOIDEA
CRABRONIDAE
Crabroninae
    477   *Lestica clypeata*
Pemphredoninae
    478   *Spilomena beata*
    479   *Spilomena curruca*
    480   *Spilomena enslini*
    481   *Spilomena troglodytes*

SPHECIDAE
Sphecinae
    482   *Sphex funerarius*
ANDRENIDAE
Andreninae
    483   *Andrena agilissima*
    484   *Andrena argentata*
    485   *Andrena barbilabris*
    486   *Andrena nigroaenea*
    487   *Andrena alfkenella*
    488   *Andrena falsifica*
    489   *Andrena floricola*
    490   *Andrena minutula*
    491   *Andrena minutuloides*
    492   *Andrena semilaevis*
    493   *Andrena subopaca*
    494   *Andrena chrysosceles*
    495   *Andrena angustior*
    496   *Andrena congruens*
    497   *Andrena dorsata*
    498   *Andrena haemorrhoa*
APIDAE
Nomadinae
    499   *Nomada baccata*
    500   *Nomada castellana*
    501   *Nomada flavoguttata*
    502   *Nomada goodeniana*
    503   *Nomada ruficornis*
    504   *Nomada sheppardana*
HALICTIDAE
Halictinae
    505   *Lasioglossum albipes*
    506   *Lasioglossum calceatum*
    507   *Lasioglossum laeve*
    508   *Lasioglossum lativentre*
    509   *Lasioglossum quadrinotatum*
    510   *Sphecodes ephippius*
    511   *Sphecodes ferruginatus*
    512   *Sphecodes miniatus*
    513   *Sphecodes monilicornis*
MEGACHILIDAE
Megachilinae
    514   *Coelioxys afra*
    515   *Coelioxys brevis*
    516   *Coelioxys quadridentata*
    517   *Megachile lapponica*

DISTRIBUTION MAPS AND SPECIES PROFILES

Maps 462 to 517 show the recorded distribution of the individual species. Records are presented for three recording periods:

- \+ before 1900
- o 1900 - 1969
- • 1970 - 2014

It should be mentioned here that plus signs and open circles do not necessarily mean that the species has declined since 1900 or 1969. They may indicate that the locality has not been visited, or that the species was not looked for. In the case of the Dryinidae, which are very under-recorded, we have included in the text distributional data that does not always appear on the relevant map, perhaps because it is too vague to locate accurately.

SPECIES PROFILES

Threat statuses (for Britain only) were identified for some species in the British Red Data Book (RDB) (Shirt, 1987), in which the data sheets for aculeate Hymenoptera were compiled by G R Else and the late G M Spooner. Some of these RDB statuses were proposed for revision by Falk (1991) in a national review of scarce and threatened aculeates; such proposed changes being prefixed with a p – thus pRDB. Species with restricted distributions that failed to meet the RDB threat criteria were also listed by Falk (1991) as Notable (now referred to as Scarce). Two degrees of Notable status were recognised – Na (thought to occur in 30 or fewer 10km squares) and Nb (thought to occur in between 31 and 100 10km squares). For a full explanation of all the RDB and Notable statuses see Ball (1994). For updates to the threat status of species within the UK refer to the Joint Nature Conservation Committee (JNCC) website (http://jncc.defra.gov.uk/page-3408).

In the text of this Atlas, county names are those of the Watsonian vice-county system.

Plant names are given only in the vernacular form in the species profiles. Readers requiring scientific names should refer to the List of Plant Names at the end of the species accounts. All botanical names are as given in Stace (1991).

Map 462 *Anteon ephippiger* (Dalman, 1818)
[Dryinidae: Anteoninae]

Also known under a number of synonyms (Else, Bolton & Broad, 2016).

Distribution
Throughout England, Wales and Scotland with a few records from Ireland (including Dunkerron, Kerry; Landenstown, Kildare; Royal Canal, Kildare; Anthdown, Wicklow; Woodend, Wicklow; Tollymore Park, Down: records from O W Richards and M Olmi).

Overseas it occurs in Austria, Bulgaria, Czechoslovakia, Denmark, Finland, France, Germany, Hungary, Italy, Korea, Mongolia, Norway, Romania, Russia, Spain, Sweden, Switzerland, Turkey and Yugoslavia.

Status (in Britain only)
No status is available, but probably not threatened.

Habitat
Found in a wide range of habitats including bog, calcareous grassland, fen, lowland heathland, dry acid grassland, coastal vegetated shingle, coastal sand dunes, saltmarsh and sandy woodland.

Flight period
Mainly found during July, also June and August, rarely May and September.

Biology
Acting as a parasitoid/predator on leaf-hoppers (Hemiptera: Cicadellidae). Hosts – Deltocephalini: *Deltocephalus pulicaris* (Fallén) (Italy); Macropsini: *Macropsis* sp. (Poland); *Macrosteles laevis* (Ribaut) (Poland, Germany); *Macrosteles sexnotatus* (Fallén) (England); Athysanini: *Mocydia crocea* (Herrich-Schäffer) (England): Opsiini, *Opsius lethierryi* Wagner (Morocco); *Opsius* sp. (Spain); *Opsius stactogalus* Fieber (Italy); Paralimnini: *Psammotettix striatus* (Linnaeus) (USSR?).

Flowers visited
No data available.

Parasites
No data available.

Map compiled by: J T Burn, M E Archer, S P M Roberts.
Authors of profile: J T Burn, M E Archer.

Map 463 *Anteon gaullei* Kieffer, 1908

[Dryinidae: Anteoninae]

Also known under a number of synonyms (Else, Bolton & Broad, 2016).

Distribution
Throughout England and Wales, with a few records from Scotland and Ireland (Murlough House, Tollymore Park, Kerry; Athdown, Wicklow).

Overseas it occurs in Austria, Czechoslovakia, Denmark, Finland, France, Germany, Hungary, Holland, Italy, Norway, Sweden, Switzerland and Russia.

Status (in Britain only)
No status is available, but probably not threatened.

Habitat
Found in a wide range of habitats including bog, coastal vegetated shingle, lowland dry acid grassland, fen, lowland heathland, coastal sand dunes, sandy woodland and disused sand quarry. Can usually be found by sweeping grass.

Flight period
Mainly found during July, also June and August, rarely September.

Biology
Acting as a parasitoid/predator on leaf-hoppers (Hemiptera: Cicadellidae). Hosts – Macropsini: *Macropsis* sp. (Poland).

Flowers visited
No data available.

Parasites
No data available.

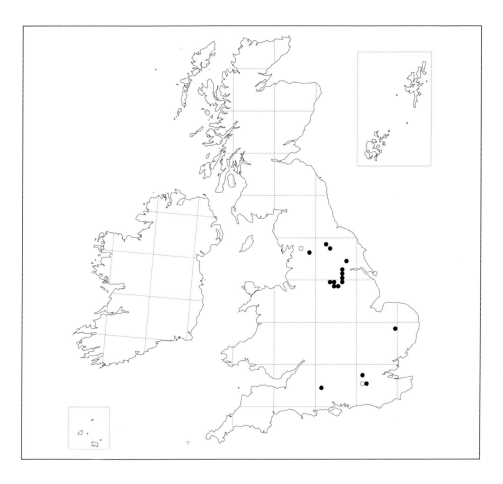

Map compiled by: J T Burn, M E Archer, S P M Roberts.
Authors of profile: J T Burn, M E Archer.

Map 464 *Anteon infectum* (Haliday, 1837)

[Dryinidae: Anteoninae]

Also known under a number of synonyms (Else, Bolton & Broad, 2016).

Distribution
Throughout England, one record from Ireland (Athdown, Wicklow) (Olmi, 1984); one record from Scotland (Dumfries) (Olmi, 1984); not recorded from Wales.

Overseas known from Austria, Denmark, Finland, France, Germany, Holland, Hungary, Italy, Japan, Norway, Romania, Sweden and Switzerland.

Status (in Britain only)
No status is available, but probably not threatened.

Habitat
Found in a wide range of habitats including lowland heathland, dry acid grassland, fen, sandy deciduous and mixed woodland, parkland, heathland, disused sand quarry, parks and hedgerows on oak.

Flight period
Mainly found during June, also May, rarely July.

Biology
Acting as a parasitoid/predator on leaf-hoppers (Hemiptera: Cicadellidae). Hosts – Iassinae: *Iassus lanio* (Linnaeus) (England, Italy); *Iassus scutellaris* (Fieber) (Italy).

Flowers visited
No data available.

Parasites
No data available.

Map compiled by: J T Burn, M E Archer, S P M Roberts.
Authors of profile: J T Burn, M E Archer.

Map 465 *Lasius alienus* (Förster, 1850)

[Formicidae: Formicinae]

Lasius alienus is a small brown to dark brownish ant. The scapes and tibia have no erect hairs. Workers can be separated from the sibling species *Lasius psammophilus* Seifert by having no or fewer hairs (0–2) between the propodeal spiracle and the metapleural gland.

Distribution
Most records are from southern and central England with a few records from Wales and Scotland. It seems to be widely distributed in Britain but absent from Ireland, although previous confusion with *Lasius psammophilus* means that it is currently under-recorded. Widespread in the Channel Islands.

There are records from Spain to Kazakhstan and in Scandinavia to 55.3° North.

Status (in Britain only)
This species is not regarded as scarce or threatened.

Habitat
It prefers warm dry habitats such as chalk grasslands. On sand and gravel substrates it is usually replaced by *Lasius psammophilus*.

Flight period
July to September.

Foraging behaviour
Tends both underground and tree-dwelling aphids and feeds on their honeydew.

Nesting biology
Nests are usually in soil and under rocks.

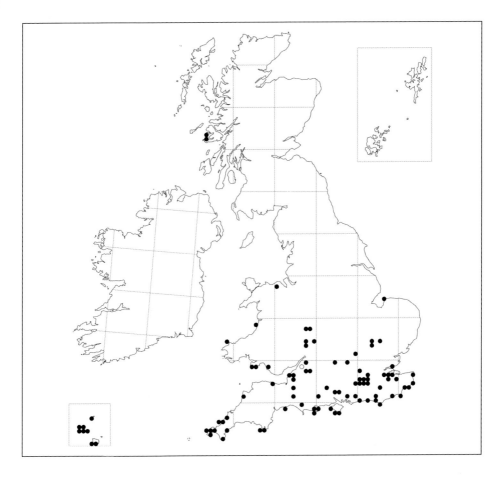

Map compiled by: M G Fox and S P M Roberts.
Author of profile: M G Fox.

Map 466 *Lasius mixtus* (Nylander, 1846)

[Formicidae: Formicinae]

Lasius mixtus workers are small and yellowish. The petiole is low with an emarginate dorsal border. There are no standing hairs on scapes or front tibiae and body hairs are shorter than *Lasius flavus* (Fabricius). Queens are brownish black with head about as broad as alitrunk. Males are brownish black with weakly dentate mandibles.

Distribution
Throughout Britain. Also recorded from the Channel Islands, but there are no recent records from Ireland.

It occurs throughout north Eurasia and Central Europe.

Status
This species is not regarded as scarce or threatened.

Habitat
This is one of the least thermophilic *Lasius* species and it inhabits pastures, sparse woodlands and grasslands.

Flight period
Alates fly in August and September. De-alate females are sometimes found wandering above ground in spring when they can take advantage of the lower aggression shown by their hosts in cold weather.

Foraging behaviour
They forage underground on small invertebrates and tend root-feeding aphids.

Nesting biology
Nests are founded by taking over a nest of *Lasius flavus* or another *Lasius* species. Nests are in the soil; sometimes under stones or among roots, alternatively mound nests of fine soil are constructed.

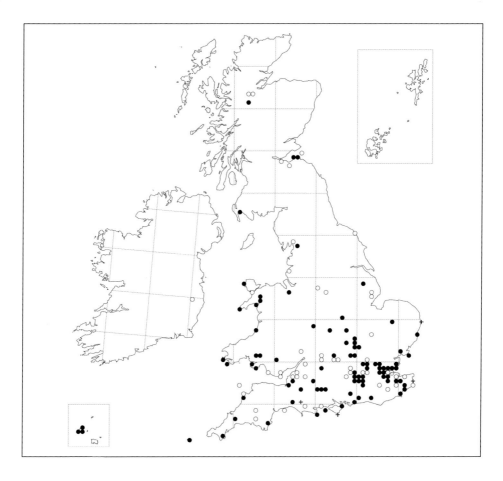

Map compiled by: M G Fox and S P M Roberts.
Author of profile: M G Fox.

Map 467 *Lasius niger* (Linnaeus, 1758)

[Formicidae: Formicinae]

Lasius niger, the common black ant, is a small, brown to dark brownish black ant that is common in gardens and, if not our commonest ant, is certainly the one people are most familiar with. The scapes and tibia have erect hairs. The clypeus has dense pubescence.

Distribution
Common in southern Britain, more local in Ireland and Scotland. There are scattered records from Ireland.

It occurs from Portugal to Baikal and from North Africa to Finland. The North American species currently regarded as *Lasius niger* has not been reviewed since Wilson (1955). Because it has not been included in the more recent studies of Palaearctic *Lasius*, the North American *Lasius niger* cannot be confidently regarded as being true *Lasius niger*.

Status (in Britain only)
This species is not regarded as scarce or threatened.

Habitat
Found in almost any dry and open area where the ground is warmed by the sun. This includes parks, gardens, roadside verges, pavements, coastal areas and brownfield sites. Not usually found in woods or other damp or shady areas. No preference for soil type is apparent.

Flight period
Alates are the familiar flying ants sometimes seen in large numbers on certain warm summer afternoons. These nuptial flights most often take place from early July to mid August.

Foraging behaviour
Workers forage boldly in the open where they are often seen running around on bare surfaces. They are aggressive and will attack other ants. They sometimes cover their foraging trails and sometimes the stems of plants where aphids are feeding with surface tunnels of earth. They scavenge and predate on small invertebrates. They also tend aphids and "milk" them for honeydew. Early in the season, when outdoor food sources are scarce, they often enter houses where they are attracted to sweet substances, and can become a nuisance. This usually stops when outdoor food sources become more plentiful.

Nesting Biology
Nests can be large and are often located under stones, paving slabs, pieces of wood, plastic or metal: in fact anything that can be warmed by the suns rays. Where there are no suitable stones etc. then large soft mound nests are sometimes found. Colonies have a single queen and from several hundred up to 10,000 workers.

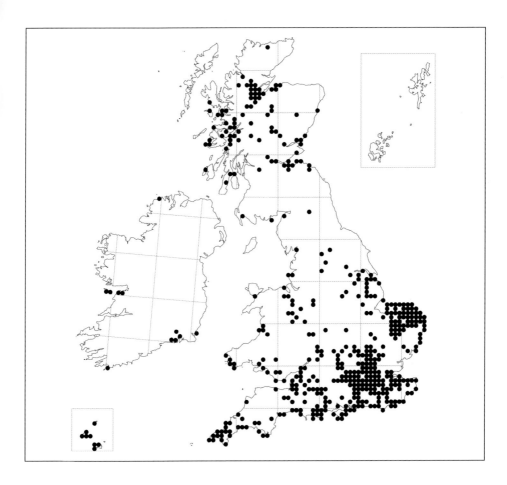

Map compiled by: M G Fox and S P M Roberts.
Author of profile: M G Fox.

Map 468 *Lasius platythorax* Seifert, 1991

[Formicidae: Formicinae]

Lasius platythorax is a small, brown to dark brownish black ant which until recently was confused with *Lasius niger* (Linnaeus). Seifert (1991) showed that it was a distinct species based on morphological differences coupled with distinctly different habitat preferences and this is now accepted by most authorities. The scapes and tibia have erect hairs. The clypeus has sparse pubescence and the hairs on the pronotum are longer than those of *Lasius niger*.

Distribution
It seems to be widely distributed in the Britain and Ireland, but previous confusion with *Lasius niger* means that it is under-recorded at this time.

Records are restricted to Europe and range from Norway to the Corsican mountains.

Status (in Britain only)
No status is available as it was only recognised subsequent to the publications of Shirt (1987) and Falk (1991). Current data suggest that this species should not be regarded as scarce or threatened.

Habitat
Found in cooler damper locations compared to its sibling species *Lasius niger*. Damp heaths and woodlands are typical habitats. Unlike *Lasius niger* it seems to avoid human habitation and is not usually found in parks or gardens. No preference for soil type is apparent.

Flight period
Seifert (2007) reports that the nuptial swarms take place on hot days some 230–130 minutes before sunset during July and August. I know of no reports of nuptial flights in Britain and anyone living close to colonies could add to our knowledge by making observations. Whether these flights coincide or not with *Lasius niger* flights would be interesting to know.

Foraging behaviour
Little has been reported on foraging behaviour. It is an aggressive species and bites readily.

Nesting Biology
Builds its nests in organic substrates, especially dead wood, with partly rotted tree stumps often being utilised. Nests can also be found in peat, leaf litter and grass tussocks but only rarely in soil. Colonies have a single queen and up to several thousands of workers.

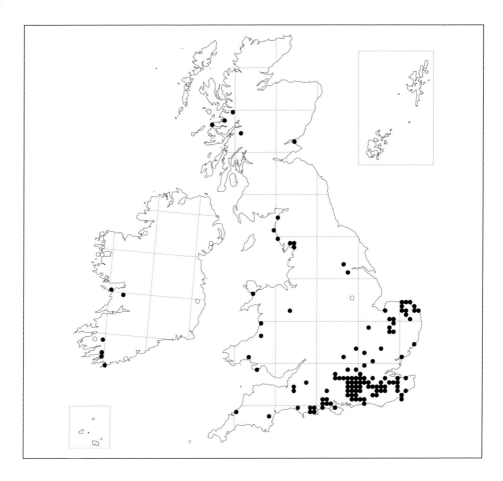

Map compiled by: M G Fox and S P M Roberts.
Author of profile: M G Fox.

Map 469 *Lasius psammophilus* Seifert, 1992
[Formicidae: Formicinae]

Lasius psammophilus is a small brown to dark brown ant which until recently was confused with *Lasius alienus* Förster. Seifert (1992) showed that it was a distinct species based on morphological differences coupled with distinctly different habitat preferences. The scapes and tibia have no erect hairs. Workers can be separated from the sibling species *Lasius alienus* by the greater number of hairs (2–5) between the propodeal spiracle and the metapleural gland.

Distribution
It is distributed in England, Wales and Ireland with a bias towards coastal sites and sandy heaths. Also occurring on the Channel Islands. Previous confusion with *Lasius alienus* means that it is probably currently under-recorded.

It is found in central Europe and in Scandinavia to 61° North.

Status (in Britain only)
This species is not regarded as scarce or threatened.

Habitat
It prefers dry grasslands and heathland over a sand or gravel substrate.

Flight period
June to September. Nuptial flights are on muggy days from 15.00 to 18.00 hours.

Foraging behaviour
Tends underground aphids and feeds on their honeydew.

Nesting biology
This species builds extensive nests, totally underground, with up to 12,000 workers in a nest.

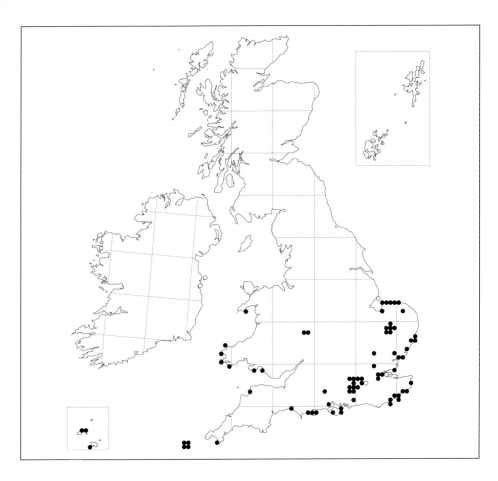

Map compiled by: M G Fox and S P M Roberts.
Author of profile: M G Fox.

Map 470 *Lasius sabularum* (Bondroit, 1918)

[Formicidae: Formicinae]

Lasius sabularum is one of the *Lasius* species with yellow workers. It is very similar to *Lasius mixtus* (Nylander), but differs in having 2–3 sub-erect hairs on the hind tibia. Males have serrated mandibles. This species has been re-established and redefined by Seifert (1988).

Distribution
Southern England and the coasts of north-west England and Wales.

It occurs across the western Palaearctic.

Status
This species is not regarded as being scarce or threatened.

Habitat
More often found in parks and gardens but also in agricultural land and meadows.

Flight period
July to October. Sometimes alates overwinter and fly in May.

Foraging behaviour
They forage underground on small invertebrates and tend root-feeding aphids.

Nesting biology
Nests are founded by taking over a nest of *Lasius niger* (Linnaeus) and probably other *Lasius* species.

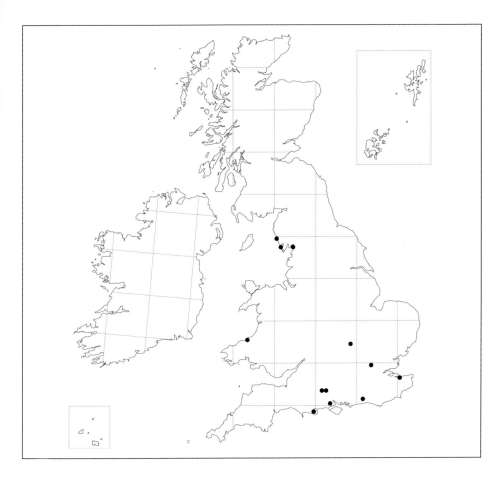

Map compiled by: G A Collins and S P M Roberts.
Author of profile: G A Collins.

Map 471 *Lasius umbratus* (Nylander, 1846)
[Formicidae: Formicinae]

Lasius umbratus workers are yellowish and rarely seen due to their underground habits. Unlike the common yellow ant *Lasius flavus* (Fabricius), *Lasius umbratus* workers have numerous erect hairs on their scapes and tibiae. Queens are reddish brown and have heads that are broader than the maximum width of the alitrunk. Males are brownish black and have denticulate mandibles.

Distribution
Throughout the Britain to central Scotland. Also recorded from Ireland.

It occurs throughout Europe.

Status (in Britain only)
This species is not regarded as scarce or threatened.

Habitat
They can be found in most areas where their host nests occur including open woodland and urban areas. They avoid very dry and very wet habitats.

Flight period
Mid August to late September.

Foraging behaviour
Workers are rarely seen above ground. They forage underground on small invertebrates and tend root-feeding aphids.

Nesting biology
It nests under boulders, in tree stumps and at the base of old trees. Nests are often carton-like structures of thin-walled chambers made of fibres and minerals bonded together with secretions and fungal hyphae. New colonies are founded by single queens invading nests of *Lasius niger* (Linnaeus), *Lasius platythorax* Seifert or *Lasius brunneus* (Latreille) and possibly *Lasius psammophilus* Seifert. In late summer queens are sometimes seen wandering over the surface of a *Lasius niger* nest carrying a dead worker as a prelude to securing adoption.

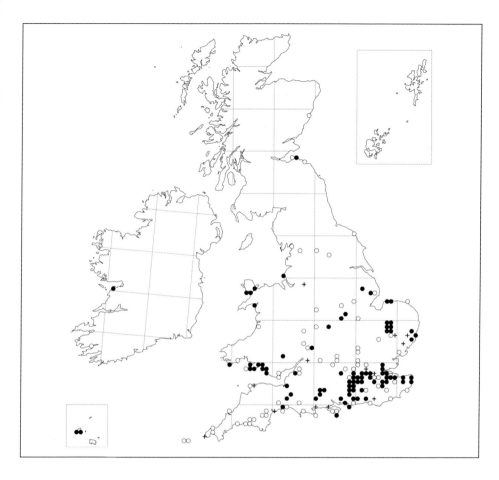

Map compiled by: M G Fox and S P M Roberts.
Author of profile: M G Fox.

Map 472 *Myrmica rubra* (Linnaeus, 1758)
[Formicidae: Myrmicinae]

Myrmica rubra is one of the common 'red' ants. Workers have antennal scapes that are long and slender with a gentle curve at the base. The area between the spines on the propodeum is smooth and shining and the spines are shorter than in other species of *Myrmica*. The petiole node has an indistinct dorsal area sloping smoothly down to its join with the postpetiole. Queens are similar to workers but larger. Male scapes are long and slender. The male tibiae and tarsi have long projecting hairs.

Distribution
Locally common all over Britain, but predominantly coastal in Ireland. Not recorded from the Channel Islands.

It occurs in almost all of Europe and Palaearctic Asia.

Status (in Britain only)
This species is not regarded as scarce or threatened.

Habitat
Prefers sunny, warm, damp habitats such as meadows and river banks but found in most open urban and agricultural areas, woodland edges, parks and gardens. Rarely found in hot, dry, sparsely vegetated areas.

Flight period
Nuptial flights usually occur during August and September.

Foraging behaviour
They feed on honeydew from aphids and scale insects and drink nectar from flowers. They are aggressive and sting readily.

Nesting biology
Nests are in the ground, in tufts of grass, under stones and in rotten wood. Colonies are usually polygynous with an average of 15 queens and a thousand workers or more.

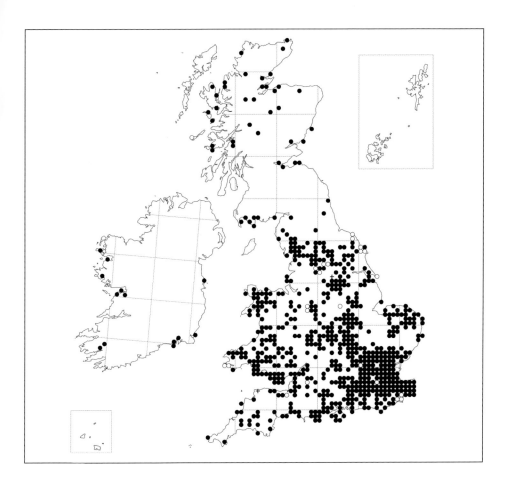

Map compiled by: M G Fox and S P M Roberts.
Author of profile: M G Fox.

Map 473 *Priocnemis confusor* Wahis, 2006

[Pompilidae: Pepsinae]

Until recently this species was known as *gracilis* Haupt, 1927 (Wahis, 2006; Collins, 2010). The male is relatively distinctive on the subgenital plate, but the female belongs to the difficult group of *Priocnemis*.

Distribution

In Britain this species is predominantly south-eastern, but with scattered records across to Wales and north to Yorkshire. Recorded from Ireland in O'Connor *et al.* (2009). Not recorded from the Channel Islands.

Overseas it occurs in central Europe (Wolf, 1972), France, Italy, Belgium, Netherlands, Germany, Austria, Denmark, Poland, Czech Republic, Slovakia and Romania (Wahis, 2011). Wiśniowski (2009) adds Finland, Sweden and Hungary, also Turkey and Siberia.

Status (in Britain only)

This species was listed as Rare (RDB3) by Shirt (1987), but revised to Nationally Notable/Nb (now known as Nationally Scarce) by Falk (1991). Current data suggest that it is scarce but not of conservation concern.

Habitat

Day (1988) describes it as a species of woodland, and more open ground, on heavy clay. In Surrey, many of the records are from sandy areas (Baldock, 2010). In the Netherlands it seems to avoid clay soils but has been recorded on sands and loess (an accumulation of wind-blown silt with sand and clay, often calcareous) (Lefeber & van Ooijen, 1988), and it also favours loess in Poland (Wiśniowski, 2009).

Flight period

Occurs most frequently in July and August, with smaller numbers in June and September.

Prey collected

There is very little data on the spider prey. A Kentish female was recorded with an immature *Clubiona* (Clubioindae). Spiders of the family Salticidae have also been recorded on the continent.

Nesting biology

There is no data. Related species nest in soil, in pre-existing cavities.

Flowers visited

There is little specific data. *Priocnemis* in general frequent white umbellifers, and *confusor* was amongst the species attracted to wild carrot lures on Surrey heathland (D Baldock, *pers. comm.*).

Map compiled by: G A Collins and S P M Roberts.
Author of profile: G A Collins.

Map 474 *Priocnemis hyalinata* (Fabricius, 1793)
[Pompilidae: Pepsinae]

Priocnemis hyalinata is one of two species of *Priocnemis* sens. str. which have an enlarged inner tooth on the tarsal claw. Females have the usual red and black pattern and are distinguished from the other species, *fennica* Haupt, by having shorter, thicker antennae. Males are black, often with some red on the second tergite, and usually with some red on the legs. They are most reliably determined by the internal genitalia which should be fully extracted.

Priocnemis fennica was only relatively recently recognised as British (Day, 1979) and older records, usually under the name *Priocnemis femoralis*, may apply to either species.

Distribution
Restricted to southern Britain, predominantly in south-east England with a few scattered records further north and in south Wales.

Overseas it occurs in Europe, except in the south, and in central Asia east to Mongolia (Wolf, 1972). Fauna Europaea (Wahis, 2011) gives most of Europe, including Fennoscandia, but excluding Spain and Portugal. The inclusion of Ireland is at odds with Day (1988) and O'Connor et al. (2009) who consider that it is *Priocnemis fennica* that occurs there.

Status (in Britain only)
This species is listed as Nationally Notable/Nb (now known as Nationally Scarce) by Falk (1991).

Habitat
Fairly open habitats, such as heathland, acid grassland and chalk downs, probably also in open areas in woods.

Flight period
A summer species, flying from late June into September, with a peak in July-August.

Prey collected
Day (1988) cites a prey record of a lycosid.

Nesting biology
No information available. Related species are fossorial, utilising natural cavities and old aculeate burrows where available.

Flowers visited
Baldock (2010) records it at wild carrot (used as a lure) and bramble.

Map compiled by: G A Collins and S P M Roberts.
Author of profile: G A Collins.

Map 475 *Priocnemis parvula* Dahlbom, 1845
[Pompilidae: Pepsinae]

A medium-sized, black and red species. In the female the clear window in the otherwise infuscate wing-tip, usually characteristic of its subgenus, is absent or poorly developed. The male subgenital plate is fairly distinctive, but must be extruded fully.

Distribution
Throughout mainland Britain, although mainly coastal in the west, and most frequent in the south-east. Not recorded from Ireland (O'Connor *et al.*, 2009), but known from the Channel Islands.

Overseas it occurs in western Europe and Fennoscandia, although absent from the Iberian peninsula and other parts of the extreme south (Wahis, 2011). Also found in central Asia and North Africa (Wolf, 1972).

Status (in Britain only)
This species is not regarded as being scarce or threatened.

Habitat
It is predominantly a species of sandy soils; heathland, acid grassland and coastal sites.

Flight period
From May to September, but more frequent in late summer.

Prey collected
Prey consists mainly of spiders in the family Lycosidae, also Thomisidae and Salticidae.

Nesting biology
No data available. Related species excavate nests in soil, often taking advantage of natural cavities and abandoned aculeate burrows.

Flowers visited
It has been recorded visiting wild carrot (Baldock, 2010; *pers. obs.*).

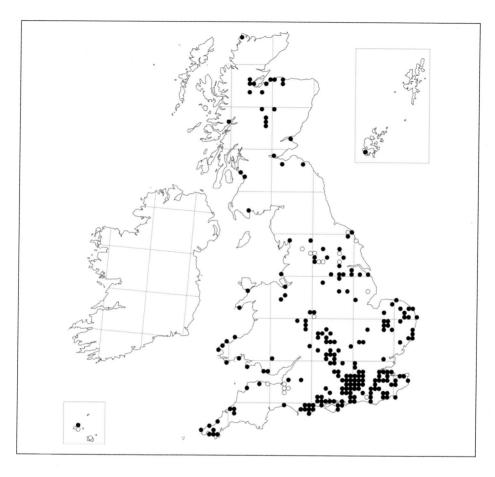

Map compiled by: G A Collins and S P M Roberts.
Author of profile: G A Collins.

Map 476 *Priocnemis pusilla* Schiødte, 1837
[Pompilidae: Pepsinae]

A medium-sized, black and red species. It belongs to the subgenus *Priocnemis* sens. str., which includes, amongst the females, some of the most difficult of the British species to identify. Typical specimens have transverse striation on the propodeum well developed, relatively short, thick antennae, and a narrow metapostnotum. Males have distinct genital plates, but must be prepared properly to appreciate this character.

Distribution
Widespread in southern and central England, also occurring in the south-west and Wales where it is mainly coastal and just reaching southern Scotland. Absent from Ireland (O'Connor et al., 2009).

Overseas it occurs in much of Europe, although absent from Norway and Sweden and parts of the western Balkans (Wahis, 2011). Also in North Africa, and northern Asia (Wolf, 1972).

Status (in Britain only)
This species is not regarded as being scarce or threatened.

Habitat
A species of open habitats on light soils; heathland, acid grassland, chalk downland and coastal sites.

Flight period
It flies from May to September, with a peak in August.

Prey collected
Prey consists of spiders in the families Salticidae and Clubionidae.

Nesting biology
Relatively poorly known, partly through confusion between similar species. Gros (1982) records the species, in France, using an abandoned burrow in which its own cell was excavated.

Flowers visited
Like most *Priocnemis* species it is fond of umbellifer flowers, and has been recorded from wild parsnip and wild carrot used as a lure (Baldock, 2010; *pers. obs.*).

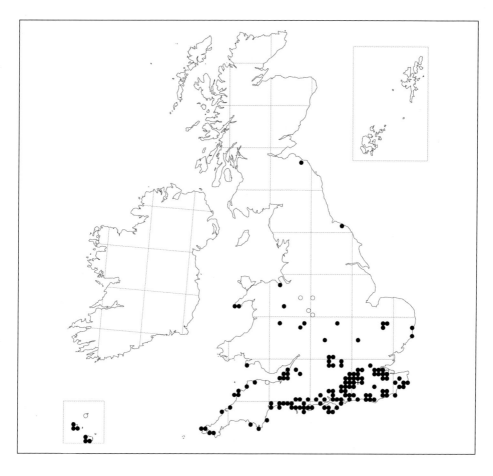

Map compiled by: G A Collins and S P M Roberts.
Author of profile: G A Collins.

Map 477 *Lestica clypeata* (Schreber, 1759)
[Crabronidae: Crabroninae]

The males of this species have a striking appearance, with a greatly elongated head constricted behind the eyes to form a distinct, almost stalk-like, 'neck'. The male's fore metatarsus is also distinctive, being drawn out into a quadrate shield-shaped process. Females are less distinctive, closely resembling females of *Ectemnius*, although the gastral terga are more strongly punctate.

Distribution
In Britain it has only ever been recorded from Weybridge in Surrey, and has not been seen there for over 150 years. The paucity of records and its early loss from Britain makes it difficult to assess its true status, but it is likely to have been a relatively short-lived colonist only, with a single record in 1848 and again in 1853 (Richards, 1980; Baldock, 2010). Given its relatively widespread occurrence in France, it is perhaps a little surprising that it has seemingly not been recorded from the Channel Islands.

It has a rather southern distribution within Europe, being very rare in Scandinavia (Lomholdt, 1984), but occurring across southern Europe, north Africa and into western and central Asia.

Status (in Britain only)
This species is listed as Extinct in Britain by Shirt (1987).

Habitat
The species is likely to be found in a variety of habitats where suitable nesting sites are found. From Carrington (1886), is is apparent that the British locality was, in general terms, an area of open heath, scrub and woodland.

Flight period
No data available from Britain. Dollfuss (1991) gives May to September for continental records.

Prey collected
Bitsch & Leclercq (1993) cite a number of European publications, where small adult Lepidoptera are listed as prey items.

Nesting biology
In mainland Europe this species nests in galleries in old timber, including large, ancient timbers and posts as well as veteran trees. Janvier (1977) describes a nest in a dead branch of an oak tree, using old galleries of xylophagous insects, in which a curved gallery led into nine individual cells. Falk (1991) notes that the nest entrances are partly closed with mud.

Flowers visited
No data available.

Parasites
No data available.

Map compiled by: A Knowles and S P M Roberts.
Author of profile: A Knowles.

Map 478 *Spilomena beata* Blüthgen, 1953
[Crabronidae: Pemphredoninae]

The small (2.5–3.5mm) species within the genus *Spilomena* probably suffer, more than any other crabronids, from under-recording. This is on account of their diminutive size, making initial discovery and capture less likely and then observation of difficult microscopic features more challenging. Rearing specimens from samples of nesting material is a useful way of searching for these wasps.

This species is not directly included within Lomholdt (1984) other than as a note to separate it from the Fennoscandian speciality *Spilomena exspectata* Valkeila, although Bitsch et al. (2001) treat *exspectata* as being synonymous with *beata*. This view is also held by Vikberg (2000). There is much taxonomic uncertainty throughout this genus.

Distribution
In Britain, it is only known from England, with the majority of records from Surrey and Kent. Elsewhere, very thinly scattered records stretch west to Devon and north to Durham. Richards (1980) describes the species as being "moderately common", although current data suggest this may be a generous summary.

It is seemingly restricted to central and northern Europe, including Germany, Switzerland, Austria, the former Czechoslovakia and Yugoslavia, Scandinavia and parts of Russia (Bitsch et al., 2001). From this it can be inferred that it is on the western edge of its range in England.

Status (in Britain only)
This species is not regarded as scarce or threatened.

Habitat
Likely to be found in a wide range of habitats where its nest sites are available, including the timbers of old houses, fences etc.

Flight period
June to September.

Prey collected
Lomholdt (1984) states that (as *Spilomena exspectata*) it collects nymphs of thrips (Thysanoptera), a view supported by other sources. This appears to be its sole source of larval food.

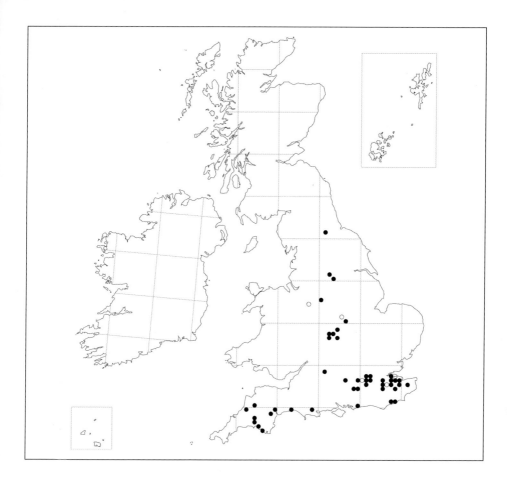

Nesting Biology

Nests in hollow plant stems or old burrows in wood. Lomholdt (1984) states that the form he treats as *Spilomena exspectata* nests in hollow or soft-pithed plant stems such as elder and bramble. Bitsch *et al.* (2001) add locations such as wood-worm borings in old windows and similar holes in the trunks of alder trees.

Flowers visited

Observations by J Leclercq cited in Bitsch *et al.* (2001) include visits to the flowers of the umbellifer *Peucedanum cervaria* (Apiaceae), a species not found in Britain.

Parasites

No data available.

Map compiled by: A Knowles and S P M Roberts.
Author of profile: A Knowles.

Map 479 *Spilomena curruca* (Dahlbom, 1843)
[Crabronidae: Pemphredoninae]

There is some taxonomic uncertainty regarding this, and other, *Spilomena* species. Here, in line with BWARS, we are currently taking the stance held by Dollfuss (1991) that this species is synonymous with *S. differens* Blüthgen. However, Lomholdt (1984), Vikberg (2000), Bitsch et al. (2001) and Else, Bolton & Broad (2016) treat the two as separate species. It would be useful to determine whether or not British specimens conform to the *curruca* sensu stricto or *differens* s.s. types, should each be promoted to true species in the future.

Distribution
Records are concentrated in south-east and central England, with only a very thin scattering of records from northern England and eastern Scotland. Richards (1980) cites Dublin, but the record is not precise enough to map.

Restricted to central and northern Europe, being on the western edge of its range in the UK. However, descriptions of European distributions are based on the species split described above. Lomholdt (1984) considers this species to be found only in Fennoscandia and Soviet Karelia (with *Spilomena differens* described as having a nondescript European distribution). Bitsch et al. (2001) say *Spilomena curruca* has been recorded from Germany, Austria, Poland and Scandinavia but NOT England, whereas *Spilomena differens* is given a broader distribution through central and northern Europe, including England.

Bitsch et al. (2001) suggest that their *Spilomena curruca* may have a boreo-montane distribution and there may be some degree of geographical/climatic separation, with *curruca* occupying more northerly and/or montane habitats than *differens*, a view supported by Vikberg (2000). This might then lead to a suggestion that the Scottish specimens mapped here might be a different "form" or species to those specimens from the south-east of England. Closer examination of specimens from northern England and Scotland would be useful in this respect.

Status (in Britain only)
This species is not regarded as scarce or threatened.

Habitat
Likely to be found in a wide range of habitats where its nest sites are available, including the timbers of old houses, fences etc. as well as old burrows of wood-worm beetles in dead trees.

Flight period
June to September.

Prey collected
Blüthgen (1953) cites a female carrying the nymph of a psyllid bug (Hemiptera: Psyllidae). Richards (1980) cites thrips, although this may be a generalisation from the genus as a whole.

Nesting Biology
Nests in hollow plant stems or old burrows in wood, including dead willow trunks.

Flowers visited
No data available.

Parasites
No data available.

Map compiled by: A Knowles and S P M Roberts.
Author of profile: A Knowles.

Map 480 *Spilomena enslini* Blüthgen, 1953
[Crabronidae: Pemphredoninae]

As one of the more recently defined taxa (Blüthgen, 1953) this species has rather more taxonomic stability than other species within the genus.

Distribution
This species is restricted to southern England, with the majority of records coming from south of the River Thames between Kent and Hampshire. There are also scattered records for Devon, Berkshire, Oxfordshire and East Suffolk. Richards (1980) additionally lists Queens County (Leix), Ireland.

This species is known across northern and central Europe.

Status (in Britain only)
This species is not regarded as scarce or threatened.

Habitat
This species might be encountered wherever suitable nesting sites are available.

Flight period
Available records range mainly from May to September, with the majority coming from June, July and August.

Prey collected
In keeping with the genus as a whole, the cells are stocked with nymphs of thrips (Thysanoptera). Lomholdt (1984) notes that *Frankliniella intonsa* (Trybom) is one such prey species.

Nesting Biology
This species appears to favour pith-filled stems of woody shrubs, such as bramble and elder. Vikberg (2000) adds raspberry and red-berried elder in Scandinavia. Despite the narrowness of these stems, Lomholdt (1984) suggests that some nests may yet be constructed with a branched pattern.

Flowers visited
No data available.

Parasites
Lomholdt (1984) lists the ichneumon wasp *Neorhacodes enslini* (Ruschka), the eupelmid wasp *Eupelmus* (as *Eupelmella*) *vesicularis* (Retzius) and the pteromalid wasps *Lonchetron fennicum* Graham and *Kaleva livida* Graham. The first three species, at least, occur in Britain. He also quotes "*Leptocryptus geniculatus* Thomson", a cryptine ichneumon wasp.

This is probably a misidentification of the species now known as *Bathythrix fragilis* (Gravenhorst) which is known to parasitise species of *Rhopalum* and *Trypoxylon* (G Broad, pers. comm.).

Map compiled by: A Knowles and S P M Roberts.
Author of profile: A Knowles.

Map 481 *Spilomena troglodytes* (Vander Linden, 1829)
[Crabronidae: Pemphredoninae]

There is some taxonomic uncertainty regarding this, and other, *Spilomena* species. Research for this Atlas suggests that BWARS initially took an outdated stance in considering this species to be distinct from *Spilomena vagans* Blüthgen (as per Richards, 1980 and Lomholdt, 1984). Dollfuss (1991), Falk (1991), Vikberg (2000) and Bitsch *et al.* (2001) all consider the two taxa to be one species: *Spilomena troglodytes*. This view is also being taken by the UK Species Inventory and implemented by the NBN Gateway. It is now accepted that *Spilomena vagans* is a synonym of *troglodytes* and so the Atlas will deal with all records under this latter name.

Distribution
The highest density of records comes from Surrey and Kent, with increasingly fewer records scattered across East Anglia, the south-west, the Midlands, northern England, and Wales. It is also recorded from Ireland by O'Connor *et al.* (2009).

This species is known across Europe eastwards to Cyprus, Turkey and Israel (Dolfuss, 1991) and also Kazakhstan and Kyrgyzstan (Bitsch *et al.*, 2001). Lomholdt (1984) (who, as an earlier author, maintained the split into the two taxa discussed above) implies a possible geographical split between the two species, with his *vagans* being found in all Nordic countries, but not common in the rest of Europe, whilst *troglodytes* was noted from only a handful of records in Fennoscandia and Denmark but widespread across the rest of the continent. This view that *Spilomena vagans* is a more Nordic beast is not supported by the stated distribution given by Richards (1980) of South Devon, Bedfordshire and Warwickshire.

Status (in Britain only)
This species is not regarded as scarce or threatened and is not listed in Falk (1991), who noted the synonymy. As a separate species, *Spilomena vagans* was considered to be Rare (RDB3) by Shirt (1987).

Habitat
This species might be encountered wherever suitable nesting sites are available.

Flight period
May to September, according to Richards (1980).

Prey collected
In keeping with the genus as a whole, the cells are stocked with nymphs of thrips (Thysanoptera). Lomholdt (1984) noted that 50–60 thrips may go to provision just one cell.

Nesting Biology
All literature sources indicate the use of old wood-worm borings in standing dead wood are the favoured nest sites, although the use of roofing thatch is also mentioned, along with abandoned tunnels in Bramble stems.

Flowers visited
No data available.

Parasites
No data available.

Map compiled by: A Knowles and S P M Roberts.
Author of profile: A Knowles.

Map 482 *Sphex funerarius* Gussakovskij, 1934
[Sphecidae: Sphecinae]

It has been shown by Menke & Pulawski (2000) that the species listed as *Sphex rufocinctus* in publications since 1975 (e.g. Lomholdt, 1984; Dolfuss, 1991; Bitsch *et al.*, 1997), and which is recorded from the Channel Islands, should be known as *funerarius* Gussakowskij. With us this species is only known from the Channel Islands and, as such, does not appear in either Richards (1980) or Falk (1991). This summary of its ecology is drawn from continental literature.

Distribution
This species is absent from mainland Britain but known from the Channel Islands, where there is a single record from Jersey.

It is known from most of Europe, the Mediterranean basin including north Africa and across Asia to China (Menke & Pulawski, 2000).

Status (in Britain only)
The Channel Islands, for a number of reasons, are excluded from the geographical coverage of the British Red Data book (Shirt, 1987) and so this species does not have a conservation status.

Habitat
Habitat components presumably comprise warm, sandy ground for nest excavation and nearby rough grassland or other herbage within which its prey is sought.

Flight period
Dollfuss (1991) gives July to September for continental examples.

Prey collected
Nymphs of grasshoppers and bush-crickets (Orthoptera), with 3–5 prey items in each cell (Bitsch *et al.*, 1997).

Nesting biology
In keeping with other members of this family, the nest generally comprises a sloping tunnel (about 15 cm long according to Lomholdt (1984)) at the end of which is one, or occasionally more, larval chambers.

Flowers visited
Bitsch *et al.* (1997) suggest that this species visits a wide range of flowers from diverse families, including Apiaceae, Asteraceae, Euphorbiaceae, Oleaceae, Rutaceae and Campanulaceae.

Parasites

Lomholdt (1984) cites the miltogrammine flies *Metopia campestris* (Fallén) and *M. argyrocephala* (Meigen) as cleptoparasites. Bitsch et al. (1997) add the non-British bombyliid fly *Thyridanthrax perspicillaris* Loew.

Map compiled by: A Knowles and S P M Roberts.
Author of profile: A Knowles.

Map 483 *Andrena agilissima* (Scopoli, 1770)
[Andrenidae: Andreninae]

A large, robust mining bee, reminiscent of *Andrena cineraria* in overall appearance but in the female lacking the distinct transverse inter-alar band of that species and the hind tibia is clad with mainly white hairs, not black. The gaster of both sexes has a conspicuous blue cast; wings strongly infuscated.

Distribution
This attractive bee is absent from mainland Britain and known only from the Channel Islands, where it has been reported from Jersey, Guernsey and Alderney (though there are no recent records from the latter island). It is described as being very common in Guernsey (C David, *pers. comm.*).

This is a central and southern European species, ranging from Holland and Poland south to Spain, the Balearics, Sardinia, Sicily, Malta, Corsica, and east to Czechoslovakia. It is widely distributed in North Africa, from Morocco to Libya.

Status (in Britain only)
The Channel Islands, for a number of reasons, are excluded from the geographical coverage of the British Red Data book (Shirt, 1987). It is not considered to be scarce or threatened there.

Habitat
In Guernsey the bee is widespread but particularly so on coastal grassland, including dunes. These are sites where there is a preponderance of yellow-flowered *Brassica* – the main pollen and nectar sources (C David, *pers. comm.*).

Flight period
Univoltine; mid May to early August.

Pollen collected
Polylectic but with a strong bias for Brassicaceae.

Nesting biology
Females apparently favour vertical surfaces in which to excavate their nests. There are reports of these being encountered in soft rock cliffs, steep slopes and the mortar joints of old walls (Kocourek, 1966; Westrich, 1989; C David, *pers. comm.*). Nests have been found as small aggregations and several females have been observed sharing a common nest entrance (Westrich 1989). The species probably overwinters as an adult in its natal cell (Stöckhert, 1933).

Flowers visited
Richards (1979) and C David (*pers. comm.*) report the following flower records: creeping buttercup, thrift, cabbage and radish.

Parasites
None reported from the Channel Islands. However, elsewhere in Europe, *Nomada fulvicornis* Fabricius has been listed as a cleptoparasite of this *Andrena* (Stöckhert, 1933; Westrich, 1989).

Map compiled by: G R Else and S P M Roberts.
Author of profile: G R Else.

Map 484 *Andrena argentata* Smith, 1844
[Andrenidae: Andreninae]

This small to medium-sized *Andrena* is strongly associated with loose, dry sandy soils in open, heathy situations. The females are quite colourful when fresh, but soon become dowdy. The males, which race over the surface of loose sand in the sun, wear out even more quickly and a bright silver-grey insect rapidly becomes a dull browny-black with few obvious hairs. It is closely related to the similar, but larger and spring-flying, *Andrena barbilabris* (Kirby). Although not restricted to visiting ericaceous flowers, the heathland flowering period suits this species very well, providing ample floral resources close to suitable nesting habitat.

Distribution
This species occurs in England and has a strongly south-eastern distribution, but has increased its range recently. There are also records from the Channel Islands.

Despite its very southerly distribution in Britain, it is a northern and central European species.

Status (in Britain only)
This species is listed in Falk (1991) as Nationally Notable/Na (now known as Nationally Scarce).

Habitat
Associated with dry, warm habitats on sandy soils, most often ericaceous heathland. It has also been found in less typical locations in West Sussex, such as in a woodland clearing, although still within a sandy area.

Flight period
July to August.

Pollen collected
Widely polylectic (Westrich, 1989).

Nesting biology
Small nesting aggregations occur in loose sandy soil in sunny places, often using quite well-churned paths. The females allow the burrow to collapse behind them as they enter and leave, possibly hiding the entrance. The actual nest is made in the firmer sand underneath, so excessive soil disturbance is damaging to the nests.

Flowers visited
A range of flowers, including umbellifers, heathers and tormentil.

Parasites
Nomada baccata (Smith) is the special cleptoparasite of this bee.

Map compiled by: M Edwards and S P M Roberts.
Author of profile: M Edwards.

Map 485 *Andrena barbilabris* (Kirby, 1802)

[Andrenidae: Andreninae]

This medium-sized, spring-flying *Andrena* is strongly associated with loose, dry sandy soils. The females are quite colourful when fresh, but soon become dowdy. The males, which race over the surface of loose sand in the sun, wear out even more quickly and a bright silver-grey insect rapidly becomes a dull browny-black with few obvious hairs. It is closely related to the similar, but smaller and summer-flying, *Andrena argentata* Smith.

Distribution
It is found throughout the British Isles, although scarcer towards the north, also recorded from the Channel Islands and scattered localities in Ireland.

It is widely distributed in northern and central Europe.

Status (in Britain only)
This species is not regarded as scarce or threatened.

Habitat
Strongly associated with light, sandy soils, but widespread on these.

Flight period
March to June.

Pollen collected
Very widely polylectic.

Nesting biology
Nests in the ground, forming small aggregations; usually in patches of loose sand although they also nest between the paving stones in my sandy garden. The females allow the burrow to collapse behind them as they enter and leave, possibly hiding the entrance. The actual nest is made in the firmer sand underneath, so excessive soil disturbance is damaging to the nests.

Flowers visited
There are flower-visiting records for a wide range of spring-flowering shrubs as well as herbaceous species such as dandelion.

Parasites
Unusually for an *Andrena*, its special cleptoparasitic bee is a *Sphecodes – pellucidus* Smith. One is often alerted to its probable presence by the *Sphecodes* flying over loose sand and digging into it as it searches for nest burrows of the *Andrena*.

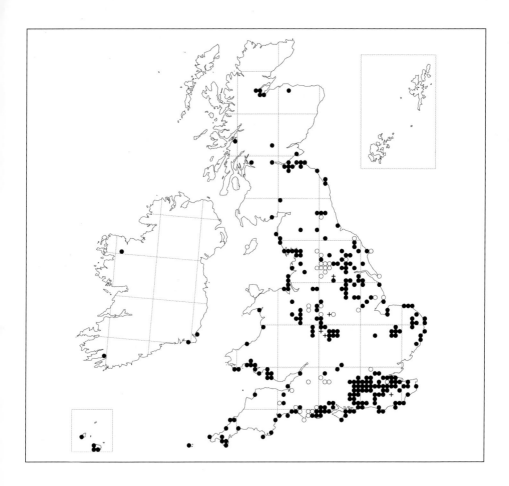

Map compiled by: M Edwards and S P M Roberts.
Author of profile: M Edwards.

Map 486 *Andrena nigroaenea* (Kirby, 1802)
[Andrenidae: Andreninae]

One of the larger *Andrena* species, with a generally dark brown abdomen and contrasting orange hairs on the hind legs and thorax. This species is one of the first to emerge in the spring, males often flying rapidly over areas of bare ground or sitting on dandelion flowers. Since the mid-1990s however, an increasing number of confirmed records of freshly emerged specimens in July point to the presence of a partial (at least) second generation. Whether this has become possible with the increase in duration of higher temperatures for a longer period in our summer is a moot point. Westrich (1989) does not note this happening in southern Germany.

Distribution
Over most of Britain, predominantly coastal in the west and extreme north. There are also scattered records for Ireland.

It is widely distributed in Europe.

Status (in Britain only)
This species is not regarded as scarce or threatened.

Habitat
It occurs in a wide range of habitats.

Flight period
April to June, with a small second generation in July and August.

Pollen collected
A widely polylectic species.

Nesting biology
Nests singly in short turf and bare ground.

Flowers visited
It visits a wide variety of flowers for nectar.

Parasites
Nomada goodeniana (Kirby) is a cleptoparasite on this species.

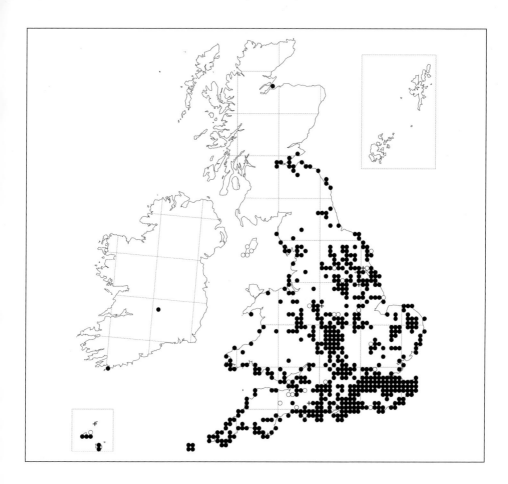

Map compiled by: M Edwards and S P M Roberts.
Author of profile: M Edwards.

Map 487 *Andrena alfkenella* Perkins, 1914
[Andrenidae: Andreninae]

This bee is one of eleven British species in the subgenus *Micrandrena*. These are very small species (body length 5–7 mm) that have an entirely black body integument relieved only by feeble transverse lateral bands of silvery hairs on tergites 2–4. Several species present difficult challenges to identification, as this often requires an appreciation of very subtle differences in surface microsculpture on the metasomal tergites and puncture density on the mesonotum.

Distribution
Southern England and East Anglia. Also recorded from the Channel Islands. It is rarely common in any one site and females are far more often encountered than males. The summer brood is generally more often seen than the spring brood. The bee is almost certainly under-recorded (for example, recent visits to chalk grasslands often produced the species in numbers) and, on account of its small size, is easily overlooked.

In mainland Europe the species occurs from southern Sweden to southern Iberia, east to Turkey (Stöckhert, 1933; Gusenleitner & Schwarz, 2002). It has also been recorded from North Africa (Morocco) (Warncke, 1981).

Status (in Britain only)
The species is listed as Rare (RDB3) by Shirt (1987) and Falk (1991). However, the species is not regarded as threatened.

Habitat
Dry, well drained soils, particularly chalk grassland, occasionally heaths, dunes, commons and cliffs both coastal and inland.

Flight period
Bivoltine. The spring brood flies from late April to early June, the summer one from July to early September. There are slight morphological differences between both sexes between the spring and autumn generations. Perkins (1914) described each of these as a distinct species: *Andrena moricella* (spring) and *alfkenella* (summer). Later (Perkins, 1919) he realised that they belonged to just a single species, *alfkenella*, the name *moricella* becoming a junior synonym of *alfkenella*.

Pollen collected
Broadly polylectic, visiting species in the families Apiaceae, Brassicaceae, Rosaceae and Scrophulariaceae (Westrich, 1989). However, in southern England the bee is particularly associated with Apiaceae flowers.

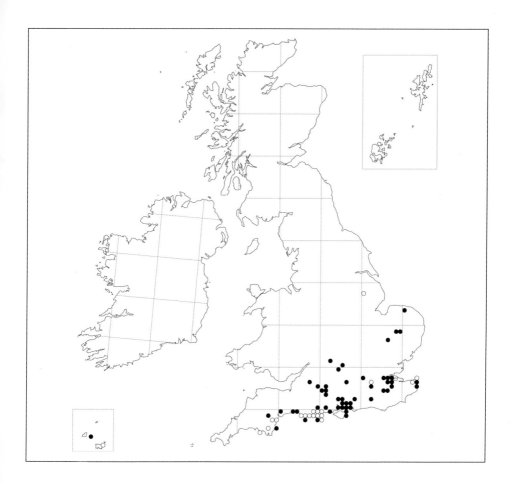

Nesting biology
Little known. In mainland Europe the species has been reported to nest in moderately sloping sand and loam banks (Kocourek, 1966).

Flowers visited
Spring brood: a cabbage, a cinquefoil, blackthorn, a speedwell, a knapweed, a cat's-ear, daisy, a mayweed. Summer brood: wild angelica, wild parsnip, hogweed, upright hedge-parsley, wild carrot.

Parasites
An unidentified *Stylops* (Strepsiptera) species has been found affecting a few females of this bee in south Devon (Spooner *in* Stidston, 1951).

Map compiled by: G R Else & S P M Roberts.
Author of profile: G R Else.

Map 488 *Andrena falsifica* Perkins, 1914
[Andrenidae: Andreninae]

Distribution
Southern England; there is a single record from North-east Yorkshire (Whitby, 13 May 1924 (Keighly Museum) (Archer, 1998)). This small bee is very local, rare and easily overlooked.

The range extends throughout Europe and west Asia; southern Scandinavia to Spain and Italy and east to Transcaucasus (Gusenleitner & Schwarz, 2002).

Status (in Britain only)
This species is listed as Nationally Notable/Na (now known as Nationally Scarce) (Falk, 1991).

Habitat
Imperfectly known though many records are from heaths and moors.

Flight period
Apparently univoltine; March or April to late June and early July.

Pollen collected
Probably polylectic (Westrich, 1989) but sources are not known.

Nesting biology
Details not known although it has been reported to nest solitarily (Kocourek, 1966; Dylewska, 1987; Westrich, 1989).

Flowers visited
Cinquefoil, wild strawberry, a speedwell, daisy.

Parasites
Nomada flavoguttata (Kirby) has been recorded as a cleptoparasite of this species. Individuals are occasionally stylopised by an unknown *Stylops* species (Strepsiptera).

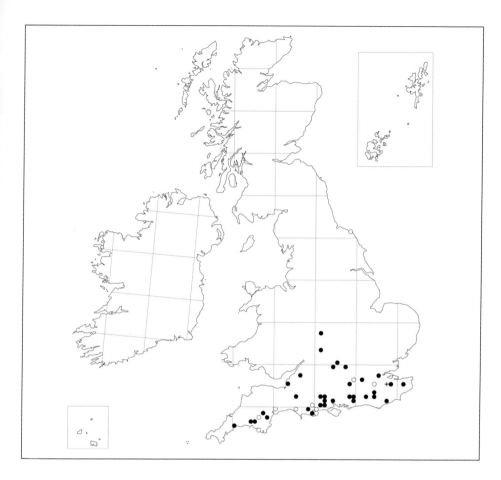

Map compiled by: G R Else & S P M Roberts.
Author of profile: G R Else.

Map 489 *Andrena floricola* Eversmann, 1852
[Andrenidae: Andreninae]

Distribution
On 11 May 1939, E Ernest collected a female of this bee at Princes Risborough, Buckinghamshire (Yarrow & Guichard, 1941). This remains the only record for the bee in Britain. The specimen is in the collection of the Natural History Museum, London.

This bee occurs throughout much of central and southern Europe (Dylewska, 1987) but is uncommon. There are apparently no records from Fennoscandia, Spain, Italy and the Mediterranean islands (Gusenleitner & Schwarz, 2002).

Status (in Britain only)
The species is listed as Endangered (RDB1) in the British Red Data Book (Shirt, 1987) and Falk (1991). However, serious consideration should be given as to whether this bee has ever been a resident in Britain.

Habitat
Not known.

Flight period
Bivoltine; the spring brood flies in April and May, the summer one in July and August.

Pollen collected
Polylectic with a strong preference for Brassicaceae and Apiaceae (Westrich, 1989).

Nesting biology
In mainland Europe the species nests solitarily in sandy and loamy soils (Kocourek, 1966; Dylewska, 1987).

Flowers visited
In mainland Europe the bee visits the following species. Spring brood: shepherd's-purse, hoary cress, rape, turnip, charlock, wild radish, a cinquefoil, a speedwell, a dandelion. Summer brood: hoary alison, shepherd's-purse, rape, turnip, charlock, wild radish, a cinquefoil, wild carrot, sheep's-bit. The British record was probably collected from a willow flower (Yarrow & Guichard, 1941).

Parasites
No data available.

Map compiled by: G R Else & S P M Roberts.
Author of profile: G R Else.

Map 490 *Andrena minutula* (Kirby, 1802)

[Andrenidae: Andreninae]

This small bee exhibits seasonal dimorphism. Following a review of the British *Micrandrena*, Perkins (1914) considered that individuals of the spring generation were a distinct species, *Andrena parvula* (Kirby). He knew the summer generation as *Andrena minutula* (Kirby). The differences between the broods were based on very subtle differences in surface microsculpture and punctation. The name *parvula* was later demoted to that of a junior synonym of *minutula* (Perkins, 1919). This is a very common species but is easily overlooked and sometimes misidentified as the closely related species *Andrena minutuloides* Perkins and *Andrena semilaevis* Pérez.

Distribution

This is perhaps the most frequently encountered species of the subgenus *Micrandrena* in the British Isles. The bee is widely distributed throughout Britain from the south coast of England northwards to southern Scotland (Kirkcudbrightshire). The species was recorded for the first time in the Isle of Man as recently as 2009 (S M Crellin, *pers. comm.*). It is sporadic in Ireland, with records from Kilkenny, Carlow, Wicklow and Dublin (Stelfox, 1927). On the Channel Islands it occurs on Jersey, Herm and Guernsey (Richards, 1979; Archer, 1996); also Sark (Beavis, 2000).

The species is very widely distributed in the Palaearctic region, the range extending from southern Fennoscandia (though very rare in Finland) south to Spain and North Africa (Morocco to Tunisia), and east to Turkey and the former USSR (Gusenleitner & Schwarz, 2002). It is additionally known from Pakistan (Quetta), China and Japan (Tadauchi, 1985). Gusenleitner & Schwarz (2002) recognise four subspecies of which only the nominate one occurs in the British Isles and Channel Islands.

Status (in Britain only)

This species is not regarded as scarce or threatened.

Habitat

Generally distributed, occurring for example in open woodland, grassland, coastal sites and in gardens.

Flight period

Bivoltine; the spring brood flies from mid March to early June, the summer one from late June or early July to the end of September.

Pollen collected

The species is broadly polylectic, visiting species in numerous families (Westrich, 1989).

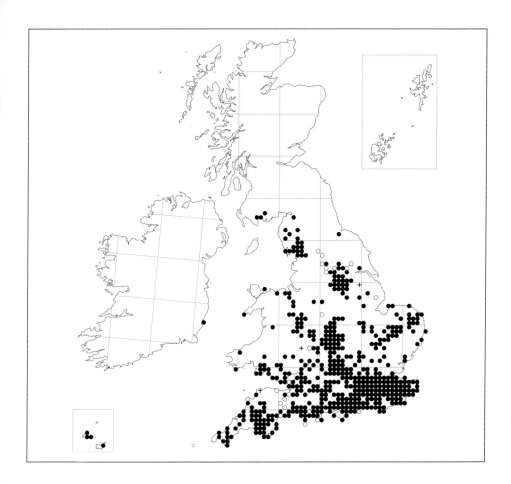

Nesting biology
Despite the local abundance of this bee, its nests are not often found. In mainland Europe it is reported to nest solitarily (Dylewska, 1987) and the same behaviour may exist in the Britain and the Channel Islands.

Flowers visited
Both broods visit the flowers of many species.

Parasites
Several authors regard this bee as being the host of the common cleptoparasitic bee *Nomada flavoguttata* (Kirby) (e.g. Perkins, 1919; Westrich, 1989). Individual bees are sometimes stylopised by *Stylops spreta* Perkins (Strepsiptera).

Map compiled by: G R Else & S P M Roberts.
Author of profile: G R Else.

Map 491 *Andrena minutuloides* Perkins, 1914
[Andrenidae: Andreninae]

This very small bee is a close relative of *Andrena minutula* (Kirby) and can be easily misidentified as such by the novice. The species is bivoltine. Perkins (1914) formerly treated the individuals of these broods as distinct species: *Andrena parvuloides* (spring) and *Andrena minutuloides* (summer). Later he relegated the name *parvuloides* to that of a junior synonym of *minutuloides* (Perkins, 1919).

Distribution
The species is confined to southern England. Possibly not present in the Channel Islands; Richards (1979) only cites a doubtful record from Guernsey. This is generally regarded as a rare and very local bee but, occasionally, can be numerous where found. The summer brood is usually much more numerous than the spring one.

In the Palaearctic the range includes southern Fennoscandia south to Iberia, Morocco, Italy, Greece and Turkey (Gusenleitner & Schwarz, 2002); also known from Daghestan.

Status (in Britain only)
Falk (1991) lists the species as Nationally Notable/Na (now known as Nationally Scarce).

Habitat
The species is usually associated with well-drained soils such as sandy heaths and commons and especially calcareous grasslands.

Flight period
Generally considered to be bivoltine, with both a spring generation (mid April to June) and a summer one (mid July to early September). However, most records of this species are confined to the summer months (post June). Specimens of the spring brood are very rarely encountered both in the field and in collections. For example, the profile author regularly recorded over several years large numbers of the summer brood near Winchester, Hampshire, but despite searches he never found specimens of the bee in the spring months.

Pollens collected
Broadly polylectic, visiting flowers in the families Aceraceae, Apiaceae, Asteraceae, Brassicaceae, Rosaceae and Scrophulariaceae (Westrich, 1989).

Nesting biology
Nests are rarely found and there is apparently no such record from Britain. In mainland Europe Kocourek (1966) and Dylewska (1987) describe the species as nesting solitarily.

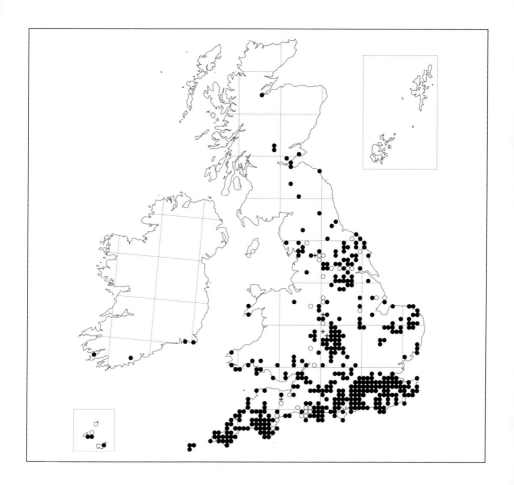

Flowers visited
Spring brood: cabbage, speedwell and daisy. Summer brood: hogweed and yarrow. In Britain the species is strongly associated with Apiaceae, such as wild parsnip, hogweed and wild carrot.

Parasites
Nests are possibly subjected to attacks by the cleptoparasitic Nomada flavoguttata (Kirby). This association is known in Germany (Westrich, 1989) and may occur in Britain too.

Map compiled by: G R Else & S P M Roberts.
Author of profile: G R Else.

Map 492 *Andrena semilaevis* Pérez, 1903
[Andrenidae: Andreninae]

This species was formerly known by the name *Andrena saundersella* (Perkins, 1914), now treated as a synonym of *semilaevis* Pérez.

Distribution
The species occurs throughout much of the British Isles. In Britain it is found from the south coast of England north to the Isle of Man and northern Scotland (Easterness). It is similarly widespread in Ireland, reaching Down in the north. In the Channel Islands it has been reported from Alderney, Herm, Guernsey, Jersey (Richards, 1979) and Sark (Beavis, 2000). A common bee that is easily overlooked on account of its small size.

This is a northern and central European species, ranging from central Fennoscandia south to the Pyrenees, Yugoslavia and Ukraine (Gusenleitner & Schwarz, 2002). It is apparently absent from the Mediterranean Basin.

Status (in Britain only)
This species is not regarded as scarce or threatened.

Habitat
Generally distributed, occurring for example in open woodland, grassland and coastal sites.

Flight period
Usually credited as being univoltine, with a peak in numbers in late spring (May). However, there are many records of specimens being recorded during the summer (as late as August) so it is possible that in some locations the species is bivoltine, with a smaller summer brood.

Pollens collected
Polylectic. Westrich (1989) lists the families Apiaceae, Asteraceae and Scrophulariaceae as being pollen sources.

Nesting biology
Not known in detail. Stelfox (1927), when describing the habits of this bee in an Irish context, simply states that it nests in dry sunny slopes and banks. Kocourek (1966) reports that in eastern Europe the species nests solitarily in loamy soil.

Flowers visited
The bee visits numerous flower species but is perhaps most closely associated with speedwell and umbellifers.

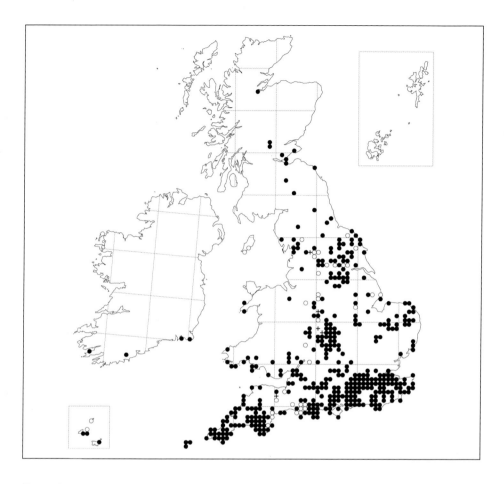

Parasites

Nests are attacked by the cleptoparasite *Nomada flavoguttata* (Kirby) (Perkins, 1919; Kocourek, 1966). Some individuals are stylopised by *Stylops spreta* Perkins (Strepsiptera).

Map compiled by: G R Else & S P M Roberts.
Author of profile: G R Else.

Map 493 *Andrena subopaca* Nylander, 1848
[Andrenidae: Andreninae]

Distribution
Throughout much of Britain and Ireland, though apparently absent from the Channel Islands. The bee is often locally common but is readily confused with its close relative *Andrena minutula* (Kirby).

The species has a Eurasian range, occurring from western Europe to Japan and Kamchatka (Tadauchi, 1985; Gusenleitner & Schwarz, 2002).

Status (in Britain only)
This species is not regarded as scarce or threatened.

Habitat
The species occurs in various habitats but is perhaps most closely associated with open deciduous woodland.

Flight period
Generally regarded as univoltine, flying from April to June. However, there are a few authentic summer records from June to September. These may represent a partial second brood.

Pollen collected
Polylectic, foraging from species in the families Caryophyllaceae, Liliaceae, Rosaceae and Scrophulariaceae (Westrich, 1989).

Nesting biology
Nests are rarely encountered. Kocourek (1966) and Westrich (1989) describe the species as nesting in small aggregations in sparsely grassed banks.

Flowers visited
A wide range of species have been recorded: buttercup, lesser stitchwort, a willow, a cabbage, bilberry, tormentil, trailing tormentil, a strawberry, hawthorn, a spurge, small-flowered crane's-bill, a water-dropwort, germander speedwell, sheep's-bit, a dandelion, daisy, oxeye daisy, a squill.

Parasites
Nomada flavoguttata (Kirby) is a cleptoparasite of this species. Individuals are occasionally stylopised by *Stylops spreta* Perkins (Strepsiptera).

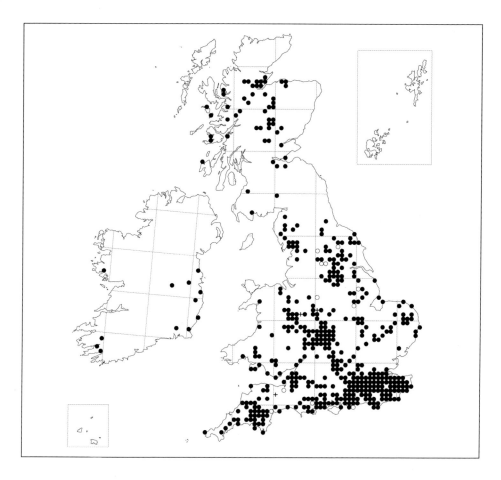

Map compiled by: G R Else & S P M Roberts.
Author of profile: G R Else.

Map 494 *Andrena chrysosceles* (Kirby, 1802)
[Andrenidae: Andreninae]

The females of this medium-sized *Andrena* can be found very commonly in late spring, often visiting the flowers of early umbellifers. Typical specimens are fairly easily recognised in the field: the male has a yellow clypeus with two small black dots at the sides, the female has a shining black abdomen with thin white hair lines at the apices of the segments and a tuft of brown-gold hairs at the tip. The scopa on the hind leg is a bright golden colour. Unfortunately, this happy state of affairs is often upset by the high frequency of stylopised specimens, where all manner of mixtures of male and female characters may occur. Confusion is most likely with female *Andrena fulvago* (Christ). However, that species flies a little later, from mid June onwards, and visits composites only.

Distribution
Found throughout England, though with surprisingly few records in the south-west, and Wales. There is one record in Scotland and none for Ireland or the Channel Islands.

It is widespread in Southern and Central Europe.

Status (in Britain only)
This species is not regarded as scarce or threatened.

Habitat
Occurs in a range of habitats, but especially at the edges of woodland and in woodland clearings on clay soils.

Flight period
Late March to early June.

Pollen collected
Widely polylectic.

Nesting biology
A solitarily nesting species.

Flowers visited
A wide range of late-spring flowers.

Parasites
This species is often parasitised by *Stylops hammella* Perkins (Strepsiptera).

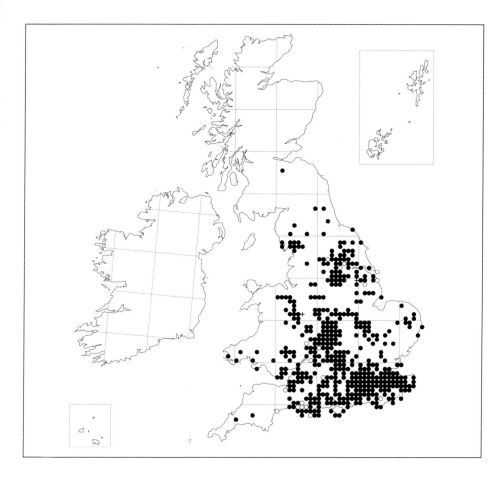

Map compiled by: M Edwards and S P M Roberts.
Author of profile: M Edwards.

Map 495 *Andrena angustior* (Kirby, 1802)

[Andrenidae: Andreninae]

The females of this medium-sized *Andrena* are very like those of *Andrena bicolor* Fabricius, but can be found in May, between the main flight periods of that species. Careful examination of the clypeus of female *Andrean angustior* will show a faintly impressed vertical line, not present in *bicolor*. Males of *Andrena angustior* seem to be generally scarce, the reason is not known.

Distribution
Found throughout England and Wales, though apparently absent from East Anglia. There are few records in Scotland or Ireland.

It is widespread in Europe.

Status (in Britain only)
This species is not regarded as scarce or threatened.

Habitat
A range of habitats, but especially at the edges of woodlands and in woodland clearings on light soils.

Flight period.
Late April to mid June.

Pollen collected
Widely polylectic.

Nesting biology
A solitary nesting species, but has been reported in small aggregations.

Flowers visited
A wide range of late-spring flowers.

Parasites
Nomada fabriciana (Linnaeus) has been reported as an associate, but this is an association with the burrows rather than a rearing record.

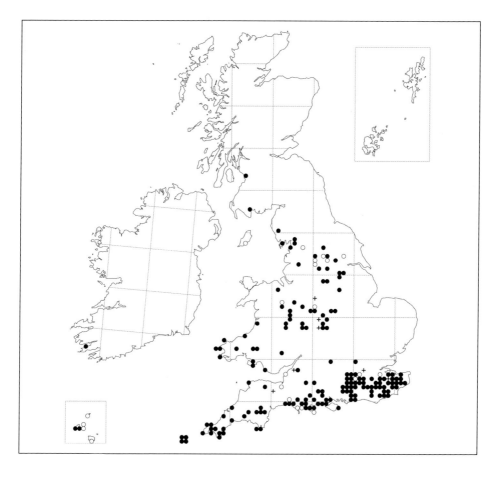

Map compiled by: M Edwards and S P M Roberts.
Author of profile: M Edwards.

Map 496 *Andrena congruens* Schmiedeknecht, 1883
[Andrenidae: Andreninae]

This medium-sized *Andrena* is very similar to the common *Andrena dorsata* (Kirby), with bright foxy hairs on the thorax and a shining black abdomen with fine lines of white hairs on the apices of the segments, only even more contrasting in appearance. It may be confirmed by careful examination, looking at the shape and form of the hairs on the hind tibia in females and the presence of a line of black hairs against the eyes in the male. It is a rather enigmatic species, being very abundant in a locality for a number of years and then apparently disappearing. I have not seen it frequently since a period in the 1990s, when it occurred in many areas in West Sussex.

Distribution
Southern England, with rather more of a central distribution than many of the southerly species, which tend to be south-easterly, also recorded in south Wales.

It is widespread in Southern and Central Europe.

Status (in Britain only)
This species is listed in Falk (1991) as Nationally Notable/Na (now known as Nationally Scarce).

Habitat
It occurs in a range of habitats with lighter soils; I have found it in sand and gravel pits as well as on the extensive chalk grasslands of Salisbury Plain.

Flight period
Bivoltine: late March to May, and July to August.

Pollen collected
Widely polylectic.

Nesting biology
A species which may nest in very large aggregations on patches of bare ground.

Flowers visited
A very wide range, including hawthorn, sallow, blackthorn, dandelion, hogweed, and wild carrot.

Parasites
None known.

Map compiled by: M Edwards and S P M Roberts.
Author of profile: M Edwards.

Map 497 *Andrena dorsata* (Kirby, 1802)
[Andrenidae: Andreninae]

A very smart-looking bee when freshly emerged, with bright foxy hairs on the thorax and a shining black abdomen with thin lines of white hairs on the apices of the segments. The males rapidly become very faded. A good character to separate the females is the form of the hind tibia which is widest at the apex, unlike the rest of the bees with this general appearance. When I first became interested in the aculeates in the 1970s this was a fairly scarce bee, but since then it has become one of the commonest andrenid bees, at least in the southern half of England.

Distribution
This species is found throughout the southern half of England, but only known from one site in Wales. Absent from Ireland but occurring on several of the Channel Islands.

It is widespread in Southern and Central Europe.

Status (in Britain only)
This species is not regarded as scarce or threatened.

Habitat
There appear to be no habitat restrictions for this species. Although there is a strong suggestion of a temperature restriction, possibly related to the need to complete two generations in a year.

Flight period
Bivoltine; March to May, and July to August.

Pollen collected
Widely polylectic.

Nesting biology
This species nests singly in short swards, and on patches of bare ground.

Flowers visited
A range of flowers are visited for nectar.

Parasites
No parasites are confirmed for this species.

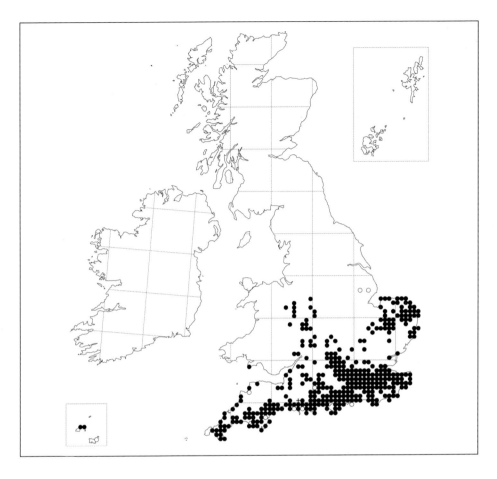

Map compiled by: M Edwards and S P M Roberts.
Author of profile: M Edwards.

Map 498 *Andrena haemorrhoa* (Fabricius, 1781)
[Andrenidae: Andreninae]

This *Andrena* is a very widespread, spring-flying species. Females are very distinctive, with a bright, foxy-coloured covering of hairs on the thorax and a similarly-coloured tuft of hairs at the tip of the abdomen, which is otherwise almost completely hairless and shining black. In common with most other *Andrena* the males are much less distinctive. However, close examination with a hand lens will reveal a dark brown spot in the middle of the otherwise orange-brown hind tibia. The thorax in fresh examples is also covered in foxy-coloured hairs, but these are much less bright than those of the females.

Distribution
This species is found throughout Britain and Ireland.

Widely distributed in Europe, although absent from parts of the extreme south such as Portugal, Spain, Sardinia and Sicily.

Status (in Britain only)
This species is not regarded as scarce or threatened.

Habitat
There appear to be no habitat restrictions for this species, beyond mountainous regions; even there it can be found in the valleys.

Flight period
March to June.

Pollen collected
Widely polylectic.

Nesting biology
This species nests singly in short swards, and along the sides of trackways.

Flowers visited
A range of flowers, probably most often found, in either sex, at those of hawthorn.

Parasites
Nomada ruficornis (Linnaeus) is the special cleptoparasite of this bee.

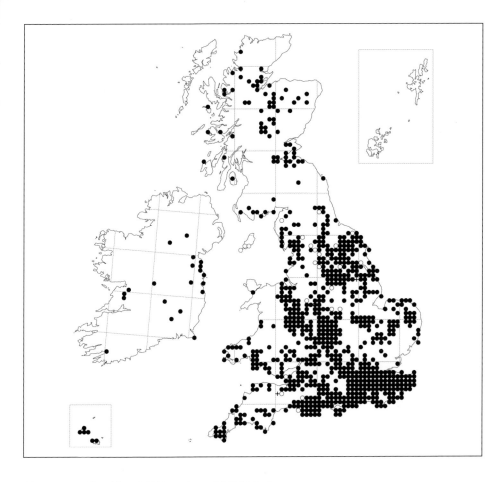

Map compiled by: M Edwards and S P M Roberts.
Author of profile: M Edwards.

Map 499 *Nomada baccata* Smith, 1844

[Apidae: Nomadinae]

One of the smaller and less frequently encountered *Nomada* species. Its unusual pattern of creamy-yellow spots on a cherry-red background is distinctive, as is its habit of flying rapidly over the surface of loose sand in July and August. Females can be picked out from the mêlée as they occasionally land, stroke the sand with their antennae and start digging. This activity signals them having found the nest-tracer scent of their host bee *Andrena argentata* Smith.

Distribution
Mainly restricted to central southern England, also recorded from East Suffolk and the Channel Islands.

It is widely distributed in northern and central Europe.

Status (in Britain only)
This species is listed in Falk (1991) as Nationally Notable/Na (now known as Nationally Scarce).

Habitat
Closely associated with sandy heathland. Although its host has been found away from heathland, it is never common in such situations and may not be sufficiently so to support a population of the *Nomada*.

Flight period
July to August.

Pollen collected
A cleptoparasitic species, it does not collect its own pollen.

Nesting biology
A specialist cleptoparasite of the bee *Andrena argentata*.

Flowers visited
Visits a wide variety of flowers.

Parasites
No data available.

Map compiled by: M Edwards and S P M Roberts.
Author of profile: M Edwards.

Map 500 *Nomada castellana* Dusmet, 1913
[Apidae: Nomadinae]

A small species with a body length of 5–7 mm. In the field it most resembles *Nomada conjungens* Herrich-Schäffer and *Nomada flavoguttata* (Kirby) in both size and coloration and could easily be passed over as being those species. The main distinguishing characters (colour and form of the labrum in the female, and shape of the intermediate flagellar segments in the male) are best viewed under a binocular microscope. The species is included in a recent key to the bees of Switzerland (Amiet *et al.*, 2007).

In the older literature this species is sometimes cited under the name *Nomada baeri* Stöckhert, 1930; a name now treated as a junior synonym of *castellana*.

Distribution
The only record is a male collected in Jersey, Channel Islands, by A C Warne (St Mary, Malaise trap, 26th June 1991 (BMNH)).

This is a little known species that is apparently widely distributed in western Europe but is sporadic and generally rare. There are, for example, records from Spain, Switzerland and Austria.

Status (in the Britain only)
The Channel Islands, for a number of reasons, are excluded from the geographical coverage of the British Red Data book (Shirt, 1987). Its status on the islands is not known.

Habitat
Not known.

Flight period
Univoltine; late April to mid July.

Pollen collected
A cleptoparasitic species, it does not collect its own pollen.

Nesting biology
In mainland Europe this species is a cleptoparasite of certain species of *Andrena* in the subgenus *Micrandrena*, possibly including *Andrena anthrisci* Blüthgen (a species not found in the British Isles) and *Andrena alfkenella* Perkins (Stöckhert, 1954; Amiet *et al.*, 2007). The host species in Jersey is not known.

Flowers visited
No records available for the Channel Islands.

Parasites
None known.

Map compiled by: G R Else and S P M Roberts.
Author of profile: G R Else.

Map 501 *Nomada flavoguttata* (Kirby, 1802)

[Apidae: Nomadinae]

Distribution
This is one of the smallest species of *Nomada* and is widely distributed throughout Britain, from southern England north to the Isle of Man and northern Scotland (Golspie, East Sutherland). In Ireland it occurs from Cork to Armagh and Down (Stelfox, 1927; Ronayne & O'Connor, 2003). Jersey is the only island in the Channel Islands where it has been reported, though not since 1903. A very common bee, although, owing to its very small size, it is easily overlooked.

The Palaearctic range encompasses southern Fennoscandia, much of central and southern Europe east to Israel, Daghestan and Japan.

Status (in Britain only)
This species is not regarded as scarce or threatened.

Habitat
Generally distributed, occurring wherever its several host species are present.

Flight period
Depending on the flight periods of its host species, it can be either univoltine or bivoltine. It can be found from the end of March to late August.

Nesting biology
The species is a cleptoparasite of certain *Andrena* species in the subgenus *Micrandrena*. Those recorded are *Andrena alfkenella* Perkins, *Andrena falsifica* Perkins, *Andrena minutula* (Kirby), *Andrena semilaevis* Pérez and *Andrena subopaca* Nylander.

Flowers visited
Many different flowers are visited for nectar.

Parasites
No data available.

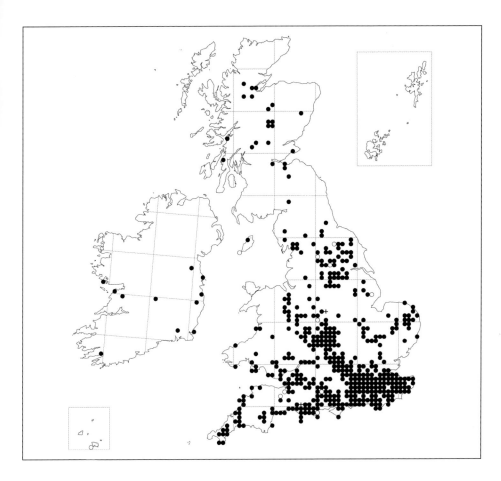

Map compiled by: G R Else & S P M Roberts.
Author of profile: G R Else.

Map 502 *Nomada goodeniana* (Kirby, 1802)

[Apidae: Nomadinae]

A very bright, black and yellow cuckoo bee and one of the largest of our *Nomada* species. In common with its hosts (*Andrena* species of the *nigroaenea*-group) this is a very widespread and frequently found species, often seen flying over short vegetation and bare ground in sunny places during the spring and, in smaller numbers, late summer. These later dates support the supposition that its hosts may also have a later flight period than has been previously recognised.

Distribution
Over most of the British Isles, although there are few Scottish records. It is also listed from Ireland (Stelfox, 1927; O'Connor *et al.*, 2009).

It is widely distributed in Europe.

Status (in Britain only)
This species is not regarded as scarce or threatened.

Habitat
It occurs in a wide range of habitats.

Flight period
April to June, with a small new generation in July and August.

Pollen collected
A cleptoparasitic species, it does not collect its own pollen.

Nesting biology
A cuckoo species, it does not make its own nest, but lays its eggs in the nests of the *Andrena nigroaenea*-group and may also be associated with other larger *Andrena*.

Flowers visited
Visits a wide variety of flowers for nectar.

Parasites
No data available.

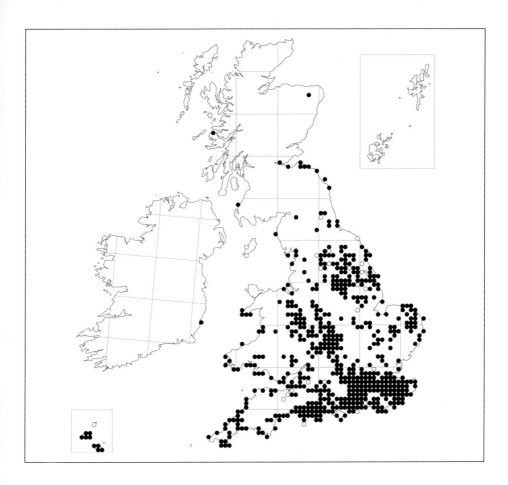

Map compiled by: M Edwards and S P M Roberts.
Author of profile: M Edwards.

Map 503 *Nomada ruficornis* (Linnaeus, 1758)

[Apidae: Nomadinae]

One of the larger brown and yellow-striped *Nomada* species. Its presence should be checked wherever its host, *Andrena haemorrhoa* (Fabricius), is present. It is very similar in appearance to the very common *Nomada flava* Panzer, but rather darker generally, though not as dark as *Nomada panzeri* Lepeletier. The presence of a clear notch in the tip of the mandible will confirm this (the only other British *Nomada* with a bifid mandible is *fabriciana* (Linnaeus), which is smaller and red). Take care when handling live females, however; female *Nomada* have quite a robust sting!

Distribution
Occurs throughout Britain and Ireland.

It is widely distributed in northern and central Europe.

Status (in Britain only)
This species is not regarded as scarce or threatened.

Habitat
As widespread as its host and in a similarly wide range of habitats.

Flight period
April to June.

Pollen collected
As a cleptoparasitic species, it does not collect its own pollen.

Nesting biology
A specialist cleptoparasite of the bee *Andrena haemorrhoa* (Fabricius).

Flowers visited
Visits a wide variety of flowers.

Parasites
No parasites are known on this species.

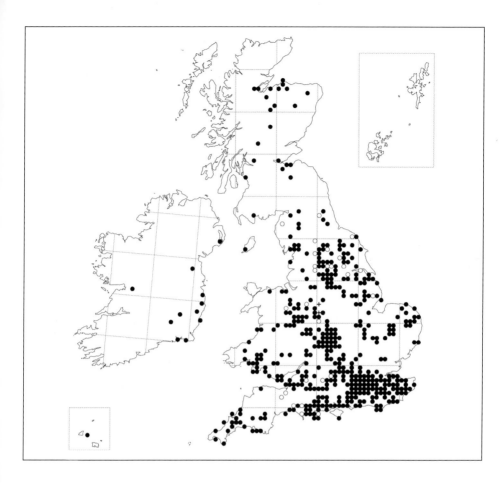

Map compiled by: M Edwards and S P M Roberts.
Author of profile: M Edwards.

Map 504 *Nomada sheppardana* (Kirby, 1802)

[Apidae: Nomadinae]

With a body length of 4–6 mm this is the smallest of the *Nomada* species found in the British Isles. The female is unusual in that the gastral tergites are devoid of yellow lateral markings of any kind. In Britain the species is unique in being entirely a cleptoparasite of small *Lasioglossum* species.

Distribution
The bee occurs throughout much of southern England from East Kent to West Cornwall, north to Salop, though apparently largely absent from eastern England north of Middlesex. It is scarce in Wales with records from Gwent (Perkins, 1918) and South Glamorgan. There are no records from northern Britain or Ireland. In the Channel Islands it is known from Guernsey (Luff, 1895; Saunders, 1902, 1903) and Jersey (Saunders, 1903).

Widely distributed in central and southern Europe, from France to Turkey; the Middle East (Israel) and North Africa (Algeria and Morocco). Also reported from Japan (Hirashima, 1982).

Status (in Britain only)
This species is not regarded as scarce or threatened.

Habitat
This bee can be expected to occur in the vicinity of the nests of its hosts in a variety of habitats.

Flight period
Apparently univoltine; end of April to late July.

Pollen collected
A cleptoparasitic species, it does not collect its own pollen.

Nesting biology
This diminutive bee is a cleptoparasite of small *Lasioglossum* species. R C L Perkins (1918) found many fully developed *Nomada sheppardana* in the burrows of a mixed nesting aggregation of *Lasioglossum parvulum* (Schenck) and *Lasioglossum nitidiusculum* (Kirby). There is speculation that other species, such as *Lasioglossum morio* (Fabricius), *Lasioglossum smeathmanellum* (Kirby) and *Lasioglossum villosulum* (Kirby) may also be hosts (Baldock, 2008; Chambers, 1949; Hallett, 1928; Smith, 1876; Perkins, 1892; Spooner, 1931).

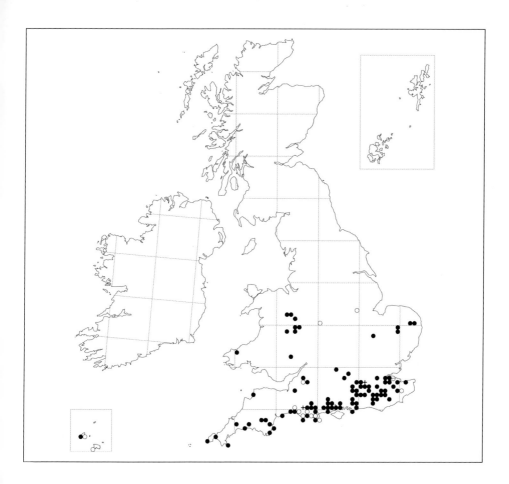

Flowers visited
An unidentified spurge is the only known flower record.

Parasites
No data available.

Map compiled by: G R Else and S P M Roberts.
Author of profile: G R Else.

Map 505 *Lasioglossum albipes* (Fabricius, 1781)
[Halictidae: Halictinae]

A very similar bee to *Lasioglossum calceatum* (Scopoli) but not as frequent as that species. The males are most easily separated by the possession of a yellow labrum, that of the male *Lasioglossum calceatum* being usually black or blackish. The females are more difficult but in *albipes* the propodeum is less sculptured posterolaterally. They are also smaller and with a longer, less-rounded face.

Distribution
Found throughout most of Britain, from the Isles of Scilly to Kent and north to parts of Scotland, but less commonly there. Also known from the Channel Islands and the Isle of Man. In Ireland, ranging north to Counties Donegal, Antrim and Down.

Abroad, a trans-Palaearctic species, ranging from the British Isles to Japan, and north to Sweden. The southern limit of its range in the west Palaearctic is in the mountains: Iberia, south to Serra de Estrela; Corsica; Italy, Monte Pollino; the high mountains of Greece south to Taygetos; north-east Turkey, but an isolated population in the south at Ankara Province (A W Ebmer, *pers. comm.*).

Status (in Britain only)
This species is not regarded as scarce or threatened.

Habitat
A woodland edge species, also found in clearings, glades and scrubby areas. At Dungeness in Kent, found on the large shingle area, sometimes commonly (*pers. obs.*).

Flight period
The female flies from mid-March to early October, whilst the male appears by early July, flying to mid-October.

Pollen collected
Polylectic but frequently visits species of buttercup for pollen.

Nesting biology
A primitively eusocial mining bee. The foundress female digs a vertical main burrow to a depth of about 15 cm. The arrangement of the cells is open to question. One report (Verhoeff, 1897) stated that the cells are sessile, i.e. without a tunnel from the main burrow, whilst in artificial conditions the cells were constructed in a cluster supported by pillars within a chamber (Plateaux-Quénu, 1989), as found in *Lasioglossum calceatum*. Between 20% and 50% of the first brood are males but although many first brood females mate they remain as workers with undeveloped ovaries. These workers prepare about 50 brood cells in which males and reproductive females are produced. These females mate and dig hibernacula by deepening the main burrow of the maternal nest.

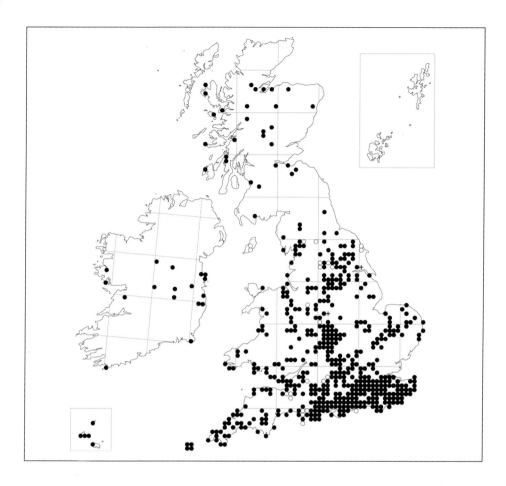

There are reports of the old foundress females surviving the winter and establishing new nests in the spring (Plateaux-Quénu, 1992).

Flowers visited
The bees visit a variety of flowers for nectar.

Parasites
There are no recorded cleptoparasites in Britain, but in continental Europe *Sphecodes monilicornis* (Kirby) is a cleptoparasite; possibly this association also exists in Britain (S Roberts, *pers. comm.*).

Map compiled by: G W Allen and S P M Roberts.
Author of profile: G W Allen.

Map 506 *Lasioglossum calceatum* (Scopoli, 1763)
[Halictidae: Halictinae]

Distribution
Found throughout Britain, from the Isles of Scilly and Kent north to the north coast of the Scottish mainland. Also found in the Channel Islands, the Isle of Man and parts of Ireland.

Found across the Palaearctic region from Britain to Japan, reaching north to northern Finland. In the south of its range, it is mostly montane, occurring up to 1,800 metres: Iberia, south to Serra Estrela; Corsica; Italy, Monte Pollino; the high mountains of Greece south to Taygetos; north-east Turkey but south to Toros, Bolkar Daglari (A W Ebmer, *pers. comm.*).

Status (in Britain only)
This species is not regarded as scarce or threatened.

Habitat
Found in most habitats, and can be abundant in more open, dry areas.

Flight period
The female is found on the wing from mid-March to early October and the male from late June to end October, at least in the south. In the northern part of the range the flight period may be partly contracted.

Pollen collected
The female is widely polylectic but often found on Asteraceae.

Nesting biology
This is a primitively eusocial mining bee in the southern part of its range. In northerly latitudes it may be solitary. The nest is constructed in short turf or other open situations in the sun, but seems not to be found in large aggregations. The nest is usually founded by one female but rarely more than one are found. Here the dominant individual, usually the largest, becomes the queen and the others then serve as workers. In the more typical situation, the lone foundress female digs the length of all the main burrow, which is almost vertical. She then constructs a short lateral tunnel some distance from the bottom, which leads to a comb-like cell cluster surrounded by an air chamber. The lone female builds 4–7 cells, in which she rears smaller workers with under-developed ovaries and sometimes one or two males. The workers take over the foraging, helping rear a brood of males and sexual females. In the south of its range, there may be two broods of workers (Plateaux-Quénu, 1992).

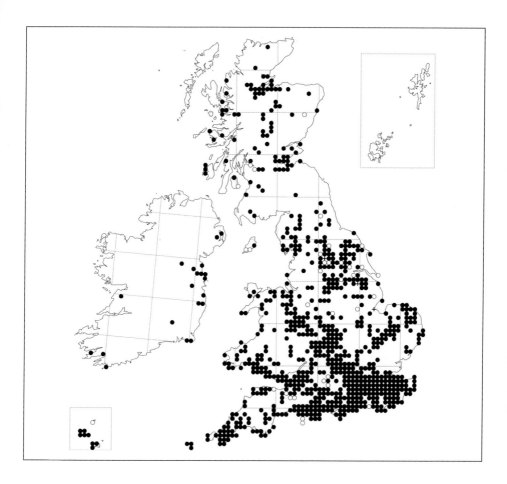

Flowers visited
The females and males will visit a wide variety of flowers for nectar.

Parasites
There are no known *Sphecodes* specifically cleptoparasitic on *Lasioglossum calceatum*, although more generalist *Sphecodes* may parasitise this species.

Map compiled by: G W Allen and S P M Roberts.
Author of profile: G W Allen.

Map 507 *Lasioglossum laeve* (Kirby, 1802)
[Halictidae: Halictinae]

Distribution
This bee is thought to be long extinct in Britain, being known only from Barham and Nacton, East Suffolk, over 200 years ago.

Abroad, it is known widely but scarcely through central and southern Europe, and western Asia.

Status (in Britain only)
Listed in the Appendix (not recorded since 1900) of both Shirt (1987) and Falk (1991).

Habitat
It is thought that the British habitats were sandy, heathy situations.

Flight period
A British flight period cannot be given but abroad females are reported from April until early August and males from August to September.

Pollen collected
British pollen sources are unknown but abroad wild carrot, chicory, dandelion, a hawkbit and devil's-bit scabious have been reported.

Nesting biology
The nesting biology has apparently not been documented, either in Britain or abroad.

Flowers visited
In addition to the above listed pollen sources, the bees have been found abroad on hoary cress, goat's-beard, Cambridge milk-parsley and a hawkweed.

Parasites
No data available.

Map compiled by: G W Allen and S P M Roberts.
Author of profile: G W Allen.

Map 508 *Lasioglossum lativentre* (Schenck, 1853)

[Halictidae: Halictinae]

A very similar bee to the closely related *Lasioglossum quadrinotatum* (Kirby); the two species are best distinguished by characters of the male genitalia.

Distribution
Widespread in southern Britain, being found from the Isles of Scilly to East Kent and north to Yorkshire and in Wales but not in Scotland or the Isle of Man. It is known from the Channel Islands but only from Jersey. Recorded from Ireland in O'Connor *et al.* (2009).

Abroad, a western Palaearctic species, found from Western Europe east to Iran, north to southern Sweden and south to Iberia and Crete; accidentally introduced into the Azores (A W Ebmer, *pers. comm.*).

Status (in Britain only)
This species is not regarded as scarce or threatened.

Habitat
A woodland edge species, sometimes found in gardens in southern Britain, also from grasslands, ruderal habitats and orchards.

Flight period
Probably univoltine. The female flies from early March to late October; the male, from mid July to mid October.

Pollen collected
British pollen sources are apparently unknown but abroad the female is polylectic.

Nesting biology
Believed to be a solitary mining bee but the nesting behaviour is apparently unrecorded.

Flowers visited
Flower visits are mainly to Asteraceae, including dandelions, but other families are also used, including Rosaceae, Ranunculaceae, Salicaceae and Ericaceae.

Parasites
The cleptoparasitic bees *Sphecodes ephippius* (Linnaeus) and *Sphecodes puncticeps* Thomson are reputed to parasitise this bee.

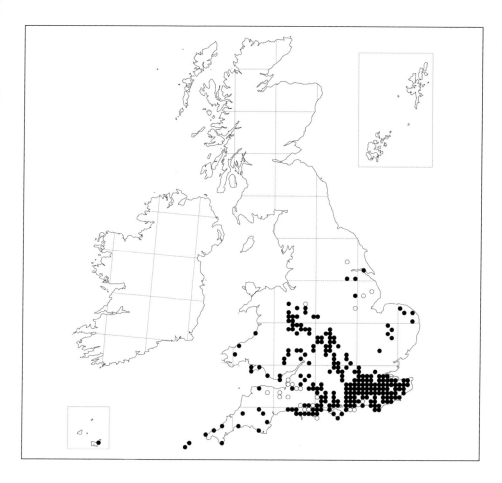

Map compiled by: G W Allen and S P M Roberts.
Author of profile: G W Allen.

Map 509 *Lasioglossum quadrinotatum* (Kirby, 1802)
[Halictidae: Halictinae]

A very similar species to the more common *Lasioglossum lativentre* (Schenck) and some records are probably due to misidentification of that species. Perkins (1922) unfortunately transposed the male genitalia characters, the most reliable way of distinguishing the two species, in his seminal paper on *Lasioglossum*.

Distribution
Found from Dorset eastwards to Kent, and north to Yorkshire and Lancashire, and found in south Wales. It is also reported from the Channel Islands.

This is a Palaearctic species, found from Britain to Kazakhstan, north to Sweden and in the south (based on males) in Italy and Greece to north-east Turkey. Reports from the south, outside these areas, are based on females which may not be correctly determined (A W Ebmer, *pers. comm.*).

Status (in Britain only)
Listed as Nationally Notable/Na (now known as Nationally Scarce) in Falk (1991).

Habitat
Found on heaths, calcareous grassland and in open woodland.

Flight period
Univoltine. The female flies from late March to late September and the male, from late July to late September.

Pollen collected
British pollen sources have not been established but in Poland, for example, it is widely polylectic, Asteraceae frequently being used.

Nesting biology
This is presumed to be a solitary mining bee, although the nesting biology is not known.

Flowers visited
The bee visits a wide range of plants in Britain, including Asteraceae, Ranunculaceae, Euphorbiacaeae, Ericaceae and Scrophulariaceae (Falk, 1991; S Roberts, *pers. comm.*).

Parasites
Sphecodes ephippius (Linnaeus) and *Sphecodes puncticeps* Thomson are believed to be cleptoparasites of this bee.

Map compiled by: G W Allen & S P M Roberts.
Author of profile: G W Allen.

Map 510 *Sphecodes ephippius* (Linnaeus, 1767)
[Halictidae: Halictinae]

Distribution
Widespread in southern Britain, being found from the Isles of Scilly and Cornwall to Kent, north to Yorkshire and Cumbria. Also known from Wales, the Isle of Man and the Channel Islands. It is not presently known from Ireland.

This is a Palaearctic species, found in south and central Europe east across Asia to Mongolia and south to North Africa (S Roberts *pers. comm.*).

Status (in Britain only)
This species is not regarded as scarce or threatened.

Habitat
Occurring in a range of habitats, such as open woodland, heaths, calcareous grassland and coastal sites.

Flight period
Univoltine; the female flies from late March to late September and the male from mid July to early October.

Pollen collected
As this bee is cleptoparasitic, no pollen is collected.

Nesting biology
The species is probably a cleptoparasite of several species of *Lasioglossum*, including the common species *calceatum* (Scopoli), *lativentre* (Schenck) and *leucozonium* (Schrank). Less common *Lasioglossum* and also *Halictus* species can be used. Occasionally *Andrena* such as *chrysosceles* (Kirby) have been implicated as hosts. It appears that the parasitic behaviour has not been observed in any detail and there are no rearing records in the literature.

Flowers visited
This *Sphecodes* visits various flowers with shallow corollas for nectar only; particularly noted are the families Asteraceae, Apiaceae and Fabaceae.

Parasites
No data available.

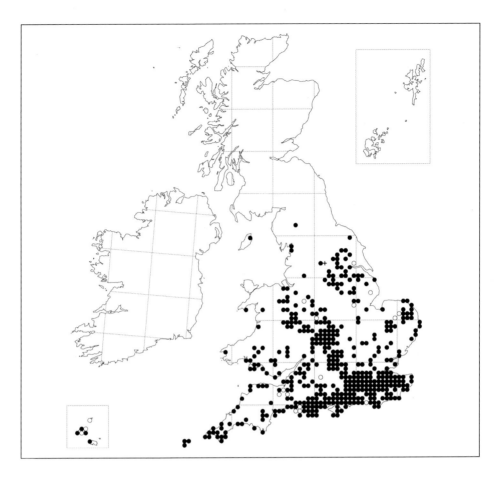

Map compiled by: G W Allen and S P M Roberts.
Author of profile: G W Allen.

Map 511 *Sphecodes ferruginatus* von Hagens, 1882
[Halictidae: Halictinae]

Distribution
A scarce species, found from Cornwall to Kent to the north of England. It is found rarely in Wales and reported from Ireland in O'Connor *et al.* (2009), but is not known from the Channel Islands.

Abroad, a Palaearctic species; known in the south from central Spain east to Greece and Azerbaijan and in the north from southern Fennoscandia east to Siberia (S Roberts *pers. comm.*).

Status (in Britain only)
Listed as Nationally Notable/Nb in Falk (1991) (now known as Nationally Scarce).

Habitat
Found in calcareous grassland, open deciduous woodland and moorland edges.

Flight period
Univoltine; the female flies from late May to mid August and the male from mid July to the end of August (Falk, 1991).

Pollen collected
As this bee is a cleptoparasite, no pollen is collected.

Nesting biology
The parasitic behaviour of this bee has apparently not been observed. It is believed to parasitise *Lasioglossum fulvicorne* (Kirby) and *Lasioglossum fratellum* (Pérez); possibly also other *Lasioglossum* such as *laticeps* (Schenck), *pauxillum* (Schenck) and *rufitarse* (Zetterstedt) (Falk, 1991).

Flowers visited
Flower visits in Britain, for nectar only, are known for a cinquefoil, fennel and wild carrot (Falk, 1991).

Parasites
No data available.

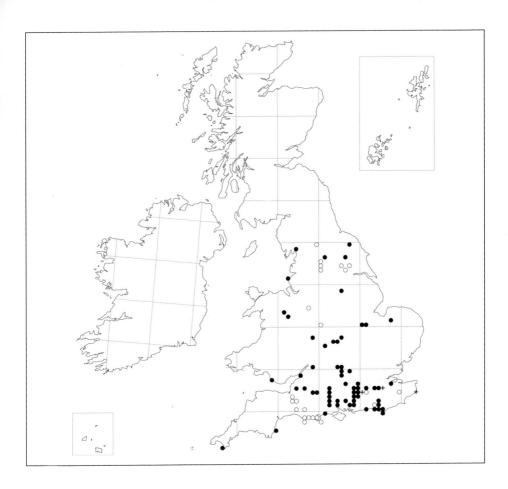

Map compiled by: G W Allen and S P M Roberts.
Author of profile: G W Allen.

Map 512 *Sphecodes miniatus* von Hagens, 1882
[Halictidae: Halictinae]

This is a very similar bee to the common *Sphecodes geoffrellus* (Kirby) and confusion, particularly in the female, most probably occurs.

Distribution
This bee, though scarce, is most frequent in south-east and eastern England. It is not known from Wales, Scotland or Ireland.

Abroad, a widespread species from Finland and Denmark, east to Samos and southern Turkey, Sultan Daglari; south to North Africa (A W Ebmer, *pers. comm.*). In central Europe it can be locally common though sporadic (S Roberts, *pers. comm.*).

Status (in Britain only)
This species was classified as Nationally Notable/Nb (now known as Nationally Scarce) by Falk (1991).

Habitat
A species of sandy habitat and frequently found on heathland.

Flight period
Univoltine, the female flying from late April to August and the male from July to September.

Pollen collected
As this bee is a cleptoparasite, no pollen is collected.

Nesting biology
This bee is most often reported to be a cleptoparasite of *Lasioglossum nitidiusculum* (Kirby) but other similar, small species such as *Lasioglossum parvulum* (Schenck) may be implicated as additional hosts. A female was taken coursing along and inspecting holes in an old ragstone wall in Kent, in which *Lasioglossum smeathmanellum* (Kirby) was nesting, another possible host (*pers. obs.*). Details of the parasitic behaviour do not appear to have been reported in the literature.

Flowers visited
Flower visits, for nectar only, are most often to plants of the daisy family, although other families are additionally used in continental Europe (S Roberts, *pers. comm.*).

Parasites
No data available.

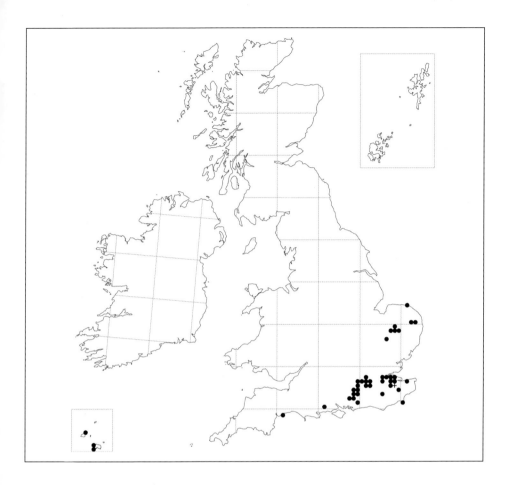

Map compiled by: G W Allen and S P M Roberts.
Author of profile: G W Allen.

Map 513 *Sphecodes monilicornis* (Kirby, 1802)
[Halictidae: Halictinae]

Distribution
Found throughout Britain, Ireland, the Isle of Man, the Isles of Scilly and the Channel Islands.

A Palaearctic species, known from Europe as far north as central Finland; also found in central Asia and North Africa (S Roberts, *pers. comm.*).

Status (in Britain only)
This species is not regarded as scarce or threatened.

Habitat
Found in a wide variety of habitats, including heaths, calcareous grassland, woodland edges and gardens.

Flight period
A univoltine species. The female flies from mid April to early October and the male, late June to early October.

Pollen collected
As this bee is a cleptoparasite, no pollen is collected.

Nesting biology
It parasitises *Halictus rubicundus* (Christ), and *Lasioglossum* species such as *malachurum* (Kirby), *albipes* (Fabricius), *calceatum* (Scopoli), *laticeps* (Schenck), *xanthopus* (Kirby) and *zonulum* (Smith).

Flowers visited
The bee visits a variety of plant families in Britain, for nectar only: Asteraceae, Apiaceae, Rosaceae, Campanulaceae and Euphorbaceae.

Parasites
No data available.

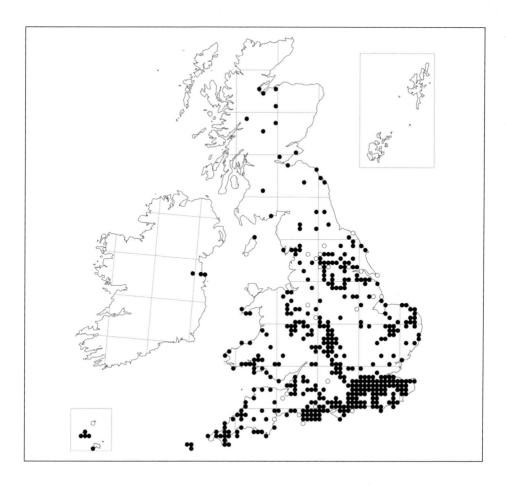

Map compiled by: G W Allen and S P M Roberts.
Author of profile: G W Allen.

Map 514 *Coelioxys afra* (Lepeletier, 1841)
[Megachilidae: Megachilinae]

A photographic test key by Rowson & Pavett (2008) is available via the BWARS website. Else and Edwards cover *Coelioxys* in their new book *Handbook of the Bees of the British Isles*, which is due for publication soon.

Distribution
There is a single British record of *Coelioxys afra* from the New Forest, South Hants (1892). This record is thought to be a vagrant or introduction. The only other records are from Guernsey in the Channel Islands (Rowson & Pavett, 2008).

In Europe it is found from Spain to Lithuania and Italy to Belgium, it has also been recorded from the Greek mainland. It is mainly recorded from European countries with coastline. Further afield, it occurs as far as Oman (Else & Edwards, *in prep.*).

Status (in Britain only)
In Britain, Shirt (1987) lists *Coelioxys afra* as Appendix (extinct) as there are no post-1900 records for the UK. The Channel Islands, for a number of reasons, are excluded from the geographical coverage of the British Red Data book.

Habitat
Little is known about the preferred habitat of *Coelioxys afra*, however in Europe it is found in open, sandy habitats (Else & Edwards, *in prep.*) which, since one of its suspected hosts is *Megachile leachella* Curtis (= *dorsalis* Pérez) (a sand dune species), is highly probable.

Flight period
The New Forest record is dated as August 1892 (Else & Edwards, *in prep.*). It is probably safe to assume that this bee has an approximately similar flight period to its known host (*Megachile leachella*), which flies from July to August, and therefore is likely to fly over a similar period and possibly slightly later than its host.

Pollen collected
No pollen is collected as it is a cleptoparasite.

Nesting biology
Coelioxys afra is a highly probable parasite of *Megachile leachella* based on the two being found in close proximity on Guernsey. In Germany, the *Magachile* is listed as being the host of *Coelioxys afra* (Westrich, 1989). The female of *afra* is the only species reported from Britain not to have a pointed sixth tergite and sternite. An egg is laid on the host's food provision or next to the host's egg (Else & Edwards, *in prep.*). Generally, *Coelioxys* larvae kill, and in some species eat, the host egg immediately on hatching (Rowson & Pavett, 2008). Pupation occurs within a cocoon spun within the host cell where the larvae overwinters as a pre-pupa, prior to final pupation, presumably in spring of the following year.

Flowers visited
There are no observations of flower visiting on the Channel Islands but in Germany *Coelioxys afra* is known to visit common bird's-foot trefoil, white melilot, viper's bugloss, wild marjoram, breckland thyme, *Teucrium* spp., greater knapweed and Irish fleabane (Westrich, 1989).

Parasites
No data available.

Map compiled by: A Jukes and S P M Roberts.
Author of profile: A Jukes.

Map 515 *Coelioxys brevis* (Eversmann, 1852)
[Megachilidae: Megachilinae]

A photographic test key by Rowson & Pavett (2008) is available via the BWARS website. Else and Edwards cover *Coelioxys* in their new book *Handbook of the Bees of the British Isles*, which is due for publication soon.

Distribution
Coelioxys brevis is absent from the British mainland and only found on Jersey in the Channel Islands.

In Europe, it is found from Spain, through France and across to Germany and Poland. It is also recorded from the Ukraine, Italy, Sardinia and Hungary. Further afield, it is recorded from North Africa. It has also been recorded eastwards to Iraq, India and Japan (Else & Edwards, *in prep.*).

Status (in Britain only)
The Channel Islands, for a number of reasons, are excluded from the geographical coverage of the British Red Data book (Shirt, 1987).

Habitat
On Jersey, *Coelioxys brevis* is only found on coastal dunes (Richards, 1979).

Flight period
Coelioxys brevis has a late flight period of mid-August to early October (Else & Edwards, *in prep.*).

Pollen collected
No pollen is collected as it is a cleptoparasite.

Nesting biology
The females of all but one British species of *Coelixoys* have a pointed sixth tergite and sternite which is apparently used to cut open the cell wall or cap of the host species. An egg is laid in this slit with at least one third protruding through into the cell or laid directly onto the host egg. Generally, *Coelioxys* larvae kill, and in some species eat, the host egg immediately on hatching. Pupation occurs within a cocoon spun within the host cell where the larvae overwinters as a pre-pupa, prior to final pupation, presumably in spring of the following year.

Flowers visited
There is no information on flower visiting by *Coelioxys brevis* on the Channel Islands. Stöckhert (1933), in Germany, cites reflexed stonecrop, viper's bugloss, breckland thyme and knapweed species as being visited by this species.

Parasites
No data available.

Map compiled by: A Jukes and S P M Roberts.
Author of profile: A Jukes.

Map 516 *Coelioxys quadridentata* (Linnaeus, 1758)
[Megachilidae: Megachilinae]

A photographic test key by Rowson and Pavett is available via the BWARS website. Else and Edwards cover *Coelioxys* in their new book *Handbook of the Bees of the British Isles*, which is due for publication soon.

Distribution
In Britain *Coelioxys quadridentata* is restricted to England and south Wales. There are no records for Scotland or Ireland. The greatest density of records is from Dorset and Surrey with more scattered records from the south-east up towards Yorkshire. The bee is also recorded from Jersey in the Channel Islands.

On the near continent, the bee is found in many western European countries (including Scandinavia) from Finland to Spain and across to Slovenia in the east. Worldwide it is found in Asia from Turkey to Siberia.

Status (in Britain only)
Shirt (1987) and Falk (1991) both list this bee as being Rare (RDB3).

Habitat
Typical habitat includes sandy heaths and coastal dune systems (Else & Edwards, *in prep*). However, it has been found in other situations, such as flying with *Anthophora quadrimaculata* (Panzer) against a wall at a Surrey railway station (D Baker, *pers. obs. in* Baldock, 2008).

Flight Period
A univoltine species flying from mid-June to early August.

Pollen collected
As the bee is a cleptoparasite, no pollen is collected.

Nesting biology
The female of *Coelioxys quadridentata* is similar to most other *Coelioxys* species in that it has a pointed sixth tergite and sternite, designed to insert an egg into the host brood cell. Generally, *Coelioxys* larvae kill, and in some species eat, the host egg immediately on hatching. Pupation occurs within a cocoon spun within the host cell where the larva overwinters as a pre-pupa, prior to final pupation in spring of the following year. *Anthophora quadrimaculata* (Panzer) is known to be a host in Britain. M Edwards (*pers. comm.*) also reared *quadridentata* from the nest of *Anthophora furcata* (Panzer). Additional hosts reported from continental Europe include *Megachile circumcincta* (Kirby) and *Megachile willughbiella* (Kirby).

Flowers visited
The bee is reported to forage from common bird's-foot trefoil and white bryony.

Parasites
No data available.

Map complied by: A Jukes and S P M Roberts.
Author of profile: A Jukes.

Map 517 *Megachile lapponica* Thomson, 1872
[Megachilidae: Megachilinae]

Else (1999) includes this species in the notes but not the key, although it should be included in his forthcoming monograph. A key is provided in Amiet et al. (2004).

Distribution
In Britain it is only known from a single record at Weybridge, Surrey, in 1847.

On the continent it is known from Austria, Belgium, Czech Republic, France, Germany, Lithuania, Finland, Sweden, south Russia, Switzerland and Poland. In Denmark it is noted as being common. Further afield, *Megachile lapponica* is found in the east Palaearctic region.

Status (in Britain only)
Shirt (1987) and Falk (1991) list *Megachile lapponica* as Appendix (extinct) as there are no post-1900 records.

Habitat
In Germany the species is known from forest edges and clearings. It is not a heat-loving species, being found in montane and boreal forested areas of Europe.

Flight Period
A univoltine species, the bee is reported to be on the wing in Germany from mid-June to August. It can be assumed that it would have a similar flight period in Britain.

Pollen collected
There is a strong preference for rosebay willowherb and it is suggested that the bee is oligolectic on this species.

Nesting biology
As with other leaf-cutter bees, *Megachile lapponica* nests in wood of varying types and in a variety of situations from tree stumps to fence posts. Reports suggest that the leaves of rosebay willowherb are used to create the brood cells.

Flowers visited
It has been reported to forage from *Rubus* species, willowherbs and Fabaceae.

Parasites
Coelioxys inermis (Kirby) is reported to be a parasite of *Megachile lapponica*.

Map complied by: A Jukes and S P M Roberts.
Author of profile: A Jukes.

REFERENCES

Amiet, F., Herrmann, M., Müller, A. & Neumeyer, R., 2004. *Anthidium, Chelostoma, Coelioxys, Dioxys, Heriades, Lithurgus, Megachile, Osmia, Stelis.* Apidae 4. Fauna Helvetica **9**.

Amiet, F., Herrmann, M., Müller, A. & Neumeyer, R., 2007. Apidae 5. *Ammobates, Ammobatoides, Anthophora, Biastes, Ceratina, Dasypoda, Epeoloides, Epeolus, Eucera, Macropis, Melecta, Melitta, Nomada, Pasites, Tetralonia, Thyreus, Xylocopa.* Fauna Helvetica **20**.

Archer, M. E., 1996. The aculeate wasps and bees (Hymenoptera: Aculeata) of Herm and Sark, Channel Islands. *Entomologist's Gazette* **47**: 53-59.

Archer, M. E., 1998. *Threatened wasps, ants and bees (Hymenoptera: Aculeata) in Watsonian Yorkshire.* A Red Data Book. PLACE Research Centre, University College of Ripon and York St. John. Occasional Paper No. 2.

Archer, M. ed., 2005. *BWARS Members' Handbook.* Centre for Ecology and Hydrology, Huntingdon.

Baldock, D. W., 2008. *Bees of Surrey.* Surrey Wildlife Trust, Pirbright.

Baldock, D. W., 2010. *Wasps of Surrey.* Surrey Wildlife Trust, Pirbright.

Ball, S. G., 1994. The Invertebrate Site Register – objectives and achievements. *British Journal of Entomology and Natural History* **7** (Suppl. 1): 2-14.

Beavis, I. C., 2000. Aculeate bees and wasps on Sark. *Report and Transactions of the Société Guernesiaise* **24**: 714-737.

Bitsch, J. & Leclercq, J., 1993. Hyménoptères Sphecidae d'Europe Occidentale Vol. 1. Faune de France, France et Régions Limitrophes **79**: 1-325.

Bitsch, J., Barbier, Y., Gayubo, S. F., Schmidt, K., & Ohl, M., 1997. Hyménoptères Sphecidae d'Europe Occidentale Vol. 2. Faune de France, France et Régions Limitrophes **82**: 1-429.

Bitsch, J., Dollfuss, H., Bouček, Z., Schmidt, K., Schmid-Egger, C., Gayubo, S. F., Antropov, A. V. & Barbier, Y., 2001. Hyménoptères Sphecidae d'Europe Occidentale Vol. 3. Faune de France, France et Régions Limitrophes **86**: 1-459.

Blüthgen, P., 1953. Alte und neue paläarktische *Spilomena*-Arten (Hym. Sphecidae). *Opuscula Entomologica* **18**: 160-179.

Carrington, J. T., 1886. Localities for beginners, No. X – St George's Hills. *Entomologist* **19**: 226-230.

Chambers, V. H., 1949. The Hymenoptera of Bedfordshire. *Transactions of the Society for British Entomology* **9**: 197-252.

Collins, G. A., 2010. A name change in the British Pompilidae. *BWARS Newsletter* Spring 2010: **33**.

Collins, G. A., & Roy, H. E., 2012. *Provisional Atlas of the aculeate Hymenoptera of Britain and Ireland. Part 8.* Biological Records Centre, Wallingford.

Czechowski, W., Radchenko, A. & Czechowska, W., 2002. *The ants (Hymenoptera, Formicidae) of Poland.* Museum and Institute of Zoology PAS, Warsaw.

Day, M. C., 1979. Nomenclatural studies on the British Pompilidae (Hymenoptera). *Bulletin of the British Museum (Natural History). (Entomology)* **38** (1): 1-26.

Day, M. C., 1988. Spider wasps. Hymenoptera: Pompilidae. *Handbooks for the identification of British Insects* **6** part 4.

Dollfuss, H., 1991. Bestimmungsschlüssel der Grabwespen Nord- und Zentraleuropas (Hymenoptera, Sphecidae) mit speziellen Angaben zur Grabwespenfauna Österreichs. *Stapfia* **24**.

Dylewska, M., 1987. Die Gattung *Andrena* Fabricius (Andrenidae, Apoidea) in Nord- und Mitteleuropa. *Acta Zoologica Cracoviensia* **30**: 359-708.

Edwards, R. ed., 1997. *Provisional Atlas of the aculeate Hymenoptera of Britain and Ireland. Part 1*. Biological Records Centre, Huntingdon.

Edwards, R. ed., 1998. *Provisional Atlas of the aculeate Hymenoptera of Britain and Ireland. Part 2*. Biological Records Centre, Huntingdon.

Edwards, R. & Broad, G. eds., 2005. *Provisional Atlas of the aculeate Hymenoptera of Britain and Ireland. Part 5*. Biological Records Centre, Huntingdon.

Edwards, R. & Broad, G. eds., 2006. *Provisional Atlas of the aculeate Hymenoptera of Britain and Ireland. Part 6*. Biological Records Centre, Huntingdon.

Edwards, R. & Roy, H. eds., 2009. *Provisional Atlas of the aculeate Hymenoptera of Britain and Ireland. Part 7*. Biological Records Centre, Wallingford.

Edwards, R. & Telfer, M. eds., 2001. *Provisional Atlas of the aculeate Hymenoptera of Britain and Ireland. Part 3*. Biological Records Centre, Huntingdon.

Edwards, R. & Telfer, M. eds., 2002. *Provisional Atlas of the aculeate Hymenoptera of Britain and Ireland. Part 4*. Biological Records Centre, Huntingdon.

Else, G. R., 1999. Identification. Leaf-cutter bees. *British Wildlife* **10**: 388-392.

Else, G. R., Bolton, B. & Broad, G. R., 2016. Checklist of British and Irish Hymenoptera - aculeates (Apoidea, Chrysidoidea and Vespoidea). *Biodiversity Data Journal* **4**: e8050. doi: 10.3897/BDJ.4.e8050

Else, G. R. & Edwards, M. (in prep.). *Handbook of the Bees of the British Isles and the Channel Islands*. The Ray Society, London.

Falk, S., 1991. A review of the scarce and threatened bees, wasps and ants of Great Britain. *Research and Survey in Nature Conservation* **35**. NCC. Peterborough.

Gros, E., 1982. Notes sur la biologie de quelques Pompilides (1re partie). *l'Entomologiste* **38**: 193-201.

Guglielmino, A. & Olmi, M., 1997. A host-parasite catalog of world Dryinidae (Hymenoptera: Chrysidoidea). *Contributions on Entomology, International* **2**: 165-298.

Guglielmino, A. & Olmi, M., 2006. A host-parasite catalog of world Dryinidae (Hymenoptera: Chrysidoidea), first supplement. *Zootaxa* **1139**: 35-62.

Guglielmino, A. & Olmi, M., 2007. A host-parasite catalog of world Dryinidae (Hymenoptera: Chrysidoidea), second supplement. *Bollettino di Zoologia Agraria e Bachicoltura, Ser. ii* **39**: 121-129.

Gusenleitner, F. & Schwarz, M., 2002. Weltweite Checkliste der Bienengattung *Andrena* mit Bemerkungen und Ergänzungen zu paläarktischen Arten (Hymenoptera, Apidae, Andreninae, *Andrena*). *Entomofauna, Supplement* **12**: 1-280.

Hallett, H. M., 1928. The Hymenoptera Aculeata of Glamorgan. *Transactions of the Cardiff Naturalists' Society* **60**: 33-67.

Hirashima, Y., 1982. Comments on the bee fauna of Japan (Hymenoptera: Apoidea). *Entomologia Generalis* **8**: 89-97.

Janvier, H., 1977. Comportement des Crabroniens (*Hymenoptera*). Ouvrage en reprographie, édité par l'auteur. **2**: 1-272

Jervis, M. A., 1977. A new key for the identification of British species of *Aphelopus* (Hym.: Dryinidae). *Systematic Entomology* **2**: 301-303.

Jervis, M. A., 1980. Life history studies on *Aphelopus* species (Hymenoptera, Dryinidae) and *Chalarus* species (Diptera, Pipunculidae), primary parasites of typhlocybine leafhoppers (Homoptera, Cicadellidae). *Journal of Natural History* **14**: 769-780.

Kinzelbach, R. K., 1971. Morphologische Befunde an Fächerflüglern und ihre phylogenetische Bedeutung (Insects: Strepsiptera). *Zoologica* **119**: 129-256.

Kocourek, M., 1966. Prodromus der Hymenopteren der Tschechoslowakei. Part 9. Apoidea, 1. *Acta Faunistica Entomologica Musei Nationalis Prague* **12** (Supplement): 1-122.

Lefeber, Br. V. & van Ooijen, P., 1988. Verspreidingsatlas van de Nederlandse Spinnendoders (Hymenoptera: Pompilidae). *Nederlandse Faunistische Mededelingen* **4**: 1-56.

Lomholdt, O., 1984. The Sphecidae (Hymenoptera) of Fennoscandia and Denmark. *Fauna Entomologica Scandinavica* **4**.

Luff, W. A., 1895. The Aculeate Hymenoptera of Guernsey. *Report and Transactions. Guernsey Society of Natural Science* **1894**: 347-355.

Menke, A. S. & Pulawski, W. J., 2000. A Review of the *Sphex flavipennis* Species Group (Hymenoptera: Apoidea: Sphecidae: Sphecini). *Journal of Hymenoptera Research* **9**: 324-346.

O'Connor, J. P., Nash, R. & Broad, G. R., 2009. *An Annotated Checklist of the Irish Hymenoptera*. The Irish Biogeographical Society, Dublin.

Olmi, M., 1984. A revision of the Dryinidae (Hymenoptera). *Memoirs of the American Entomological Institute* **37**: 1-1913.

Olmi, M., 1989. Supplement to the revision of the world Dryinidae (Hymenoptera: Chrysidoidea). *Frustula entomologica N.S.* **12**: 109-395.

Olmi, M., 1994. The Dryinidae and Embolemidae (Hymenoptera: Chrysidoidea) of Fennoscandia and Denmark. *Fauna Entomologica Scandinavica* **30**.

Olmi, M., 1999. Hymenoptera. Dryinidae-Embolemidae. *Fauna d'Italia* **37**: 1-425.

Perkins, J. F., 1976. Hymenoptera. Bethyloidea. *Handbooks for the Identification of British Insects* **6** part 3a.

Perkins, R. C. L., 1914. Synopsis of the British forms of the *Andrena minutula* group. *Entomologist's Monthly Magazine* **50**: 71-115.

Perkins, R. C. L., 1918. *Nomada furva* K. and its hosts. *Entomologist's Monthly Magazine* **54**: 226.

Perkins, R. C. L., 1919. The British species of *Andrena* and *Nomada*. *Transactions of the Entomological Society of London* **1919**: 218-319.

Perkins, R. C. L., 1922. The British species of *Halictus* and *Sphecodes*. *Entomologist's Monthly Magazine* **58**: 46-52, 94-101, 167-174.

Perkins, R. C. L., 1943. Early spring Hymenoptera and other insects on Dartmoor in 1943. *Entomologist's Monthly Magazine* **79**: 130-132.

Perkins, V. R. *in* Witchell, C. A. & Strugnell, W. B., 1892. A list of the Hymenoptera Aculeata collected in the neighbourhood of Wootton-under-Edge. *The Fauna and Flora of Gloucestershire*. G. H. James, Stroud.

Pesenko, Y. A., Banaszak, J., Radchenko, V. G. & Cierzniak, T., 2000. *Bees of the family Halictidae (excluding* Sphecodes*) of Poland: taxonomy, ecology, bionomics.* Wydawnictwo Uczelniane Wyższej Szkoły Pedagogicznej w Bydgoszczy.

Plateaux-Quénu, C., 1989. Premières observations sur le caractère social d'*Evylaeus albipes* (F.). *Actes Colloque Insectes Sociaux* **5**: 335-344.

Plateaux-Quénu, C., 1992. Comparative biological data in two closely related eusocial species: *Evylaeus calceatus* (Scop.) and *Evylaeus albipes* (F.) (Hym., Halictinae). *Insectes Sociaux* **39**: 351-364.

Richards, O. W., 1939. The British Bethylidae (s.l.) (Hymenoptera). *Transactions of the Royal Entomological Society of London* **89**: 185-344.

Richards, O. W., 1979. The Hymenoptera Aculeata of the Channel Islands. *Report and Transactions of the Société Guernésiaise* **20**: 389-424.

Richards, O. W., 1980. Scoliodea, Vespoidea and Sphecoidea. Hymenoptera, Aculeata. *Handbooks for the Identification of British Insects* **6** part 3 (b).

Ronayne, C. & O'Connor, J. P., 2003. Distributional records of some Aculeata (Hymenoptera) collected in Ireland from 1980-2002. *Bulletin of The Irish Biogeographical Society* **27**: 227-254.

Rond, J. de (in prep.). *Anteon albidicolle*, *Gonatopus albosignatus* and *Gonozius distigmus distigmus* found along the Dutch coast and reintroduced as good species (Hymenoptera: Dryinidae, Bethylidae).

Rowson, R. & Pavett, M., 2008. *A visual identification guide of British* Coelioxys *bees*. Privately published.

Saunders, E., 1902. Hymenoptera Aculeata of Jersey, Guernsey, Alderney and St. Briac (Brittany). *Entomologist's Monthly Magazine* **38**: 140-146.

Saunders, E., 1903. Hymenoptera Aculeata in Jersey. *Entomologist's Monthly Magazine* **39**: 245-248.

Seifert, B., 1988. A revision of the European species of the ant subgenus *Chthonolasius* (Insecta, Hymenoptera, Formicidae). *Entomologische Abhandlungen Staatliches Museum für Tierkunde Dresden* **51**: 143 - 180

Seifert, B., 1991. *Lasius platythorax* n. sp., a widespread sibling species of *Lasius niger* (Hymenoptera: Formicidae). *Entomologia Generalis* **16**: 69-81.

Seifert, B., 1992. A taxonomic revision of the Palaearctic members of the ant subgenus Lasius s. str. (Hymenoptera: Formicidae). *Abhandlungen und Berichte des Naturkundemuseums Görlitz* **66**: 1-67.

Seifert, B., 2007. *Die Ameisen Mittel- und Nordeuropas*. Lutra, Klitten.

Shirt, D. B., ed., 1987. *British Red Data Books 2. Insects*. NCC, Peterborough.

Smith, F., 1876. *Catalogue of British Hymenoptera in the British Museum. 1. Andrenidae and Apidae*. Second edition. British Museum, London.

Spooner, G. M., 1931. *The bees, wasps and ants (Hymenoptera, Aculeata) of Cambridgeshire*. Cambridge Natural History Society, Cambridge.

Stace, C., 1991. *New Flora of the British Isles*. Cambridge University Press, Cambridge.

Stelfox, A. W., 1927. A list of the Hymenoptera Aculeata (sensu lato) of Ireland. *Proceedings of the Royal Irish Academy* **37** (Section B, No. 22): 201-355.

Stidston, S. T. ed., 1951. Third report of the Entomological Section. 1950. *Transactions of the Devonshire Association for the Advancement of Science, Literature and Art* **83**: 90-101.

Stöckhert, F. K., 1933. Die Bienen Frankens (Hym. Apid.). Eine ökologisch-Tiergeographische Untersuchung. *Deutschen Entomologischen Zeitschrift (Beiheft)* **132**: 1-294.

Stöckhert, F. K., 1954. Fauna Apoideorum Germaniae. *Abhandlungen der Bayerischen Akademie der Wissenschaften* **65**: 1-87.

Tadauchi, O., 1985. Synopsis of *Andrena* (*Micrandrena*) of Japan. *Journal of the Faculty of Agriculture, Kyushu University* **30**: 59-76, 77-94.

Verhoeff, C., 1897. Zur Lebensgeschichte der Gattung *Halictus* (Anthophila), insbesondere einer Übergangsform zu socialen Bienen. *Zoologischer Anzeiger* (Leipzig) **20**: 369-393.

Vikberg, V. V., 2000. A re-evaluation of five European species of *Spilomena* with a key to European species and relevance to the fauna of North Europe, especially Finland (Hymenoptera: Pemphredonidae). *Entomologica Fennica* **11**: 35-55.

Wahis, R., 2006. Mise à jour du Catalogue systématique des Hyménoptères Pompilides de la région ouest-européenne. Additions et Corrections. *Notes Fauniques de Gembloux* **59**: 31-36.

Wahis, R., 2011. Fauna Europaea: Pompilidae. Fauna Europaea version 2.4, http://www.faunaeur.org.

Waloff, N., 1975. The parasitoids of the nymphal and adult stages of leaf-hoppers (Auchenorrhyncha: Homoptera) of acidic grassland. *Transactions of the Royal Entomological Society of London* **126**: 637-686.

Warncke, K., 1981. Die Bienen des Klagenfurter Beckens (Hymenoptera, Apidae). *Carinthia II* **171**: 275-348.

Westrich, P., 1989. *Die Wildbienen Baden-Württemburgs. Band 2*. Eugen Ulmer, Stuttgart.

Wilson, E. O., 1955. A monographic revision of the ant genus *Lasius*. *Bulletin of the Museum of Comparative Zoology* **113**: 1-201.

Wiśniowski, B., 2009. *Spider-hunting wasps (Hymenoptera: Pompilidae) of Poland.* Ojców National Park, Poland.
Wolf, H., 1972. Hymenoptera Pompilidae. *Insecta Helvetica, Fauna,* **5**. Zurich.
Yarrow, I. H. H. & Guichard, K. M., 1941. Some rare Hymenoptera Aculeata, with two species new to Britain. *Entomologist's Monthly Magazine* **77**: 2-13.

LIST OF PLANT NAMES

Following *New Flora of the British Isles* (Stace, 1991).

Common name	Scientific name	Family
Bilberry	*Vaccinium myrtillus*	Ericaceae
Blackthorn	*Prunus spinosa*	Rosaceae
Bramble	*Rubus fruticosus* agg.	Rosaceae
Breckland thyme	*Thymus serpyllum*	Lamiaceae
Buttercup	*Ranunculus* spp.	Ranunculaceae
Cabbage	*Brassica* spp.	Brassicaceae
Cambridge milk-parsley	*Selinum carvifolia*	Apiaceae
Cat's-ear	*Hypochaeris* spp.	Asteraceae
Charlock	*Sinapis arvensis*	Brassicaceae
Chicory	*Cichorium intybus*	Asteraceae
Cinquefoil	*Potentilla* spp.	Rosaceae
Common bird's-foot-trefoil	*Lotus corniculatus*	Fabaceae
Creeping buttercup	*Ranunculus repens*	Ranunculaceae
Daisy	*Bellis perennis*	Asteraceae
Dandelion	*Taraxacum officinale* agg.	Asteraceae
Devil's-bit Scabious	*Succisa pratensis*	Dipsacaceae
Fennel	*Foeniculum vulgare*	Apiaceae
Germander speedwell	*Veronica chamaedrys*	Scrophulariaceae
Goat's-beard	*Tragopogon pratensis*	Asteraceae
Greater knapweed	*Centaurea scabiosa*	Asteraceae
Hawkbit	*Leontodon* spp.	Asteraceae
Hawkweed	*Hieracium* spp.	Asteraceae
Hawthorn	*Crataegus* spp.	Rosaceae
Heather	*Calluna vulgaris*	Ericaceae
Hoary alison	*Berteroa incana*	Brassicaceae
Hoary cress	*Lepidium draba*	Brassicaceae
Hogweed	*Heracleum sphondylium*	Apiacaeae
Irish fleabane	*Inula salicina*	Asteraceae
Knapweed	*Centaurea* spp.	Asteraceae
Lesser stitchwort	*Stellaria graminea*	Caryophyllaceae
Mayweed	*Tripleurospermum, Matricaria* spp.	Asteraceae
Oxeye daisy	*Leucanthemum vulgare*	Asteraceaea
Radish	*Raphanus* spp.	Brassicaceae
Rape	*Brassica napus*	Brassicaceae
Reflexed stonecrop	*Sedum rupestre*	Crassulaceae
Rosebay willowherb	*Chamerion angustifolium*	Onagraceae
Sallow	*Salix* spp.	Salicaceae
Sheep's-bit	*Jasione montana*	Campanulaceae
Shepherd's-purse	*Capsella bursa-pastoris*	Brassicaceae
Small-flowered crane's-bill	*Geranium pusillum*	Geraniaceae

Speedwell	*Veronica* spp.	Scrophulariaceae
Spurge	*Euphorbia* spp.	Euphorbiaceae
Squill	*Scilla* spp.	Liliaceae
Thrift	*Armeria maritima*	Plumbaginaceae
Tormentil	*Potentilla erecta*	Rosaceae
Trailing tormentil	*Potentilla anglica*	Rosaceae
Turnip	*Brassica rapa*	Brassicaceae
Upright hedge-parsley	*Torilis japonica*	Apiaceae
Viper's-bugloss	*Echium vulgare*	Boraginaceae
Water-dropwort	*Oenanthe* spp.	Apiaceae
White bryony	*Bryonia dioica*	Cucurbitaceae
White melilot	*Melilotus albus*	Fabaceae
Wild angelica	*Angelica sylvestris*	Apiaceae
Wild carrot	*Daucus carota*	Apiaceae
Wild marjoram	*Origanum vulgare*	Lamiaceae
Wild parsnip	*Pastinaca sativa*	Apiaceae
Wild radish	*Raphanus raphanistrum*	Brassicaceae
Wild strawberry	*Fragaria vesca*	Rosaceae
Willow	*Salix* spp.	Salicaceae
Yarrow	*Achillea millefolium*	Asteraceae

CUMULATIVE INDEX TO SPECIES IN PARTS 1 to 9

Synonyms and misidentification names referred to in the text are listed in italics. Valid names are listed in normal typeface. Species in this part are in bold.

Species	Part	Page	
Agenioideus cinctellus	3	36	
Alysson lunicornis	3	78	
Alysson tumidus	3	86	
Ammophila pubescens	1	78	
Ammophila sabulosa	1	80	
Ancistrocerus albotricinctus	2	60	
Ancistrocerus antilope	3	54	
Ancistrocerus callosus	3	56	
Ancistrocerus claripennis	2	50	
Ancistrocerus gazella	2	52	
Ancistrocerus nigricornis	3	56	
Ancistrocerus oviventris	2	54	
Ancistrocerus parietinus	2	56	
Ancistrocerus parietum	2	58	
Ancistrocerus pictus	2	54	
Ancistrocerus quadratus	2	50	
Ancistrocerus scoticus	2	60	
Ancistrocerus trifasciatus	2	62	
Ancistrocerus trimarginatus	2	60	
Andrena agilissima	9	50	**Map 483**
Andrena alfkenella	9	58	**Map 487**
Andrena angustior	9	74	**Map 495**
Andrena apicata	4	82	
Andrena argentata	9	52	**Map 484**
Andrena barbilabris	9	54	**Map 485**
Andrena bicolor	6	62	
Andrena bimaculata	8	72	
Andrena bucephala	7	74	
Andrena carbonaria	8	74, 76	
Andrena chrysosceles	9	72	**Map 494**
Andrena cineraria	4	84	
Andrena clarkella	3	102	
Andrena coitana	5	70	
Andrena congruens	9	76	**Map 496**
Andrena denticulata	6	64	
Andrena dorsata	9	78	**Map 497**

Andrena falsifica	9	60	**Map 488**
Andrena ferox	4	86	
Andrena flavipes	4	88	
Andrena florea	3	104	
Andrena floricola	9	62	**Map 489**
Andrena fucata	5	72	
Andrena fulva	5	74	
Andrena fulvago	8	78	
Andrena fuscipes	6	66	
Andrena gravida	4	90	
Andrena haemorrhoa	9	80	**Map 498**
Andrena hattorfiana	3	106	
Andrena helvola	8	60	
Andrena humilis	8	80	
Andrena labialis	6	68	
Andrena labiata	5	76	
Andrena lapponica	5	78	
Andrena lathyri	4	92	
Andrena marginata	3	108	
Andrena minutula	9	64	**Map 490**
Andrena minutuloides	9	66	**Map 491**
Andrena nigriceps	8	70	
Andrena nigroaenea	9	56	**Map 486**
Andrena nigrospina	8	74	
Andrena niveata	8	72	
Andrena nitida	6	70	
Andrena nitidiuscula	4	94	
Andrena ovatula	8	84	
Andrena parvula	9	64	**Map 490**
Andrena parvuloides	9	66	**Map 491**
Andrena pilipes	8	76	
Andrena praecox	4	96	
Andrena proxima	7	76	
Andrena rosae	7	78	
Andrena ruficrus	8	68	
Andrena saundersella	9	68	**Map 492**
Andrena semilaevis	9	68	**Map 492**
Andrena simillima	6	72	
Andrena similis	8	86	
Andrena stragulata	7	80	
Andrena subopaca	9	70	**Map 493**
Andrena synadelpha	8	62	
Andrena tarsata	5	80	
Andrena thoracica	6	74	

Andrena tridentata	6	76	
Andrena trimmerana	8	66	
Andrena varians	8	64	
Andrena wilkella	8	88	
Anergates atratulus	4	24	
Anoplius caviventris	1	32	
Anoplius concinnus	2	42	
Anoplius infuscatus	3	40	
Anoplius nigerrimus	2	44	
Anoplius viaticus	3	42	
Anteon ephippiger	9	8	**Map 462**
Anteon fulviventre	8	10	
Anteon gaullei	9	10	**Map 463**
Anteon infectum	9	12	**Map 464**
Anteon jurineanum	8	12	
Anteon pubicorne	8	14	
Anthidium manicatum	1	110	
Anthophora bimaculata	6	94	
Anthophora furcata	6	96	
Anthophora plumipes	6	98	
Anthophora quadrimaculata	6	100	
Anthophora retusa	6	102	
Aporus unicolor	1	34	
Arachnospila anceps	6	36	
Arachnospila consobrina	6	38	
Arachnospila minutula	7	34	
Arachnospila rufa	3	38	
Arachnospila spissa	7	36	
Arachnospila trivialis	6	40	
Arachnospila wesmaeli	7	38	
Argogorytes fargei	3	90	
Argogorytes mystaceus	3	92	
Astata boops	2	66	
Astata pinguis	2	68	
Auplopus carbonarius	1	24	
Blepharipus dimidiatus	3	62	
Bombus barbutellus	6	124	
Bombus bohemicus	5	118	
Bombus campestris	6	126	
Bombus cullumanus	7	114	
Bombus distinguendus	3	124	
Bombus hortorum	7	116	
Bombus humilis	4	120	
Bombus hypnorum	7	118	

Bombus jonellus	6	118
Bombus lapidarius	5	120
Bombus lucorum	7	120
Bombus monticola	5	122
Bombus muscorum	5	124
Bombus pascuorum	5	126
Bombus pratorum	6	120
Bombus ruderarius	3	126
Bombus ruderatus	7	122
Bombus rupestris	3	128
Bombus soroeensis	6	122
Bombus subterraneus	4	122
Bombus sylvarum	3	130
Bombus sylvestris	6	128
Bombus terrestris	7	124
Bombus vestalis	5	128
Caliadurgus fasciatellus	3	32
Ceratina cyanea	1	122
Ceratophorus morio	8	54
Cerceris arenaria	1	86
Cerceris quadricincta	1	88
Cerceris quinquefasciata	1	90
Cerceris ruficornis	1	92
Cerceris rybyensis	1	94
Cerceris sabulosa	1	96
Ceropales maculata	2	48
Ceropales variegata	1	38
Chelostoma campanularum	7	98
Chelostoma florisomne	7	100
Chrysis angustula	7	10
Chrysis bicolor	6	16
Chrysis fulgida	4	18
Chrysis gracillima	6	18
Chrysis helleni	6	20
Chrysis ignita	7	12
Chrysis illigeri	6	20
Chrysis impressa	7	14
Chrysis longula	7	16
Chrysis mediata	7	18
Chrysis osmiae	4	20
Chrysis pseudobrevitarsis	7	20
Chrysis pustulosa	4	22
Chrysis ruddii	7	22
Chrysis rutiliventris	7	24

Chrysis schencki	7	26	
Chrysis viridula	2	22	
Chrysura hirsuta	4	20	
Chrysura radians	4	22	
Cleptes nitidulus	6	10	
Cleptes pallipes	6	12	
Cleptes semiauratus	6	12	
Coelioxys afra	**9**	112	**Map 514**
Coelioxys brevis	**9**	114	**Map 515**
Coelioxys conoidea	3	116	
Coelioxys elongata	8	110	
Coelioxys inermis	8	112	
Coelioxys mandibularis	8	114	
Coelioxys quadridentata	**9**	116	**Map 516**
Coelioxys rufescens	8	116	
Colletes cunicularius	1	100	
Colletes daviesanus	4	76	
Colletes floralis	1	102	
Colletes fodiens	4	78	
Colletes halophilus	1	104	
Colletes halophilus	3	94	
Colletes hederae	3	94	
Colletes marginatus	1	106	
Colletes similis	4	80	
Colletes succinctus	3	94	
Colletes succinctus	3	96	
Crabro binotatus	3	60	
Crabro cribrarius	1	62	
Crabro dimidiatus	3	62	
Crabro peltarius	1	64	
Crabro scutellatus	1	66	
Crossocerus annulipes	6	40	
Crossocerus binotatus	3	60	
Crossocerus dimidiatus	3	62	
Crossocerus distinguendus	5	44	
Crossocerus elongatulus	5	46	
Crossocerus exiguus	7	46	
Crossocerus leucostomus	7	48	
Crossocerus megacephalus	6	42	
Crossocerus palmipes	8	44	
Crossocerus podagricus	7	50	
Crossocerus pusillus	7	52	
Crossocerus quadrimaculatus	3	64	
Crossocerus styrius	7	54	

Species		
Crossocerus tarsatus	7	56
Crossocerus vagabundus	3	66
Crossocerus varus	7	52
Crossocerus walkeri	7	58
Crossocerus wesmaeli	6	44
Cryptocheilus notatus	1	26
Cuphopterus binotatus	3	60
Cuphopterus dimidiatus	3	62
Dasypoda altercator	2	106
Didineis lunicornis	3	78
Dienoplus tumidus	3	86
Diodontus insidiosus	4	52
Diodontus luperus	4	54
Diodontus minutus	4	56
Diodontus tristis	4	58
Dipogon bifasciatus	5	34
Dipogon subintermedius	5	36
Dipogon variegatus	5	42
Dolichovespula media	1	58
Dolichovespula norwegica	4	44
Dolichovespula saxonica	1	60
Dolichovespula sylvestris	4	46
Dryudella pinguis	2	68
Dufourea halictula	8	98
Dufourea minuta	8	100
Dufourea minuta	8	98
Ectemnius borealis	1	68
Ectemnius cavifrons	1	70
Ectemnius cephalotes	2	78
Ectemnius chrysostomus	2	84
Ectemnius continuus	2	80
Ectemnius dives	2	82
Ectemnius lapidarius	2	84
Ectemnius lituratus	2	86
Ectemnius nigrifrons	2	90
Ectemnius planifrons	2	90
Ectemnius quadricinctus	2	92
Ectemnius rubicola	2	88
Ectemnius ruficornis	2	90
Ectemnius saundersi	2	92
Ectemnius sexcinctus	2	92
Ectemnius zonatus	2	92
Elampus panzeri	5	10
Embolemus ruddii	1	14

Entomognathus brevis	3	68
Epeolus cruciger	4	116
Epeolus variegatus	4	118
Episyron rufipes	2	46
Eucera longicornis	6	102
Eucera nigrescens	6	104
Eucera tuberculata	6	104
Eumenes coarctatus	3	44
Euodynerus quadrifasciatus	3	46
Evagetes crassicornis	4	40
Evagetes dubius	1	28
Evagetes pectinipes	1	30
Evagetes siculus	4	42
Formica aquilonia	3	30
Formica candida	2	38
Formica cunicularia	5	24
Formica exsecta	1	18
Formica fusca	5	26
Formica lemani	5	28
Formica lugubris	4	28
Formica picea	2	38
Formica pratensis	5	30
Formica rufa	1	20
Formica rufibarbis	4	30
Formica sanguinea	5	32
Formica transkaucasica	2	38
Formicoxenus nitidulus	4	26
Goniozus claripennis	2	16
Gorytes bicinctus	3	84
Gorytes campestris	3	90
Gorytes laticinctus	3	80
Gorytes punctatus	3	88
Gorytes quadrifasciatus	3	82
Gymnomerus laevipes	3	50
Halictus confusus	7	86
Halictus eurygnathus	5	82
Halictus maculatus	5	84
Halictus rubicundus	5	86
Halictus tumulorum	7	88
Harpactus tumidus	3	86
Hedychridium ardens	2	18
Hedychridium coriaceum	3	14
Hedychridium cupreum	3	16
Hedychridium integrum	3	16

Hedychridium roseum	2	20
Hedychrum aureicolle	4	14
Hedychrum intermedium	4	16
Hedychrum niemelai	4	14
Hedychrum nobile	4	14
Hedychrum rutilans	4	16
Heriades truncorum	7	102
Holopyga amoenula	6	14
Holopyga generosa	6	14
Holopyga ovata	6	14
Homonotus sanguinolentus	1	36
Hoplisoides punctuosus	3	88
Hoplitis claviventris	2	122
Hoplitis leucomelana	2	122
Hoplitis spinulosa	7	104
Hylaeus annularis	7	66
Hylaeus brevicornis	3	98
Hylaeus communis	8	56
Hylaeus cornutus	3	100
Hylaeus gibbus	7	68
Hylaeus pectoralis	1	108
Hylaeus pictipes	7	70
Hylaeus signatus	8	58
Hylaeus spilotus	7	72
Lasioglossum albipes	**9**	94	Map 505
Lasioglossum angusticeps	4	98
Lasioglossum brevicorne	4	100
Lasioglossum calceatum	**9**	96	Map 506
Lasioglossum cupromicans	5	88
Lasioglossum fratellum	6	76
Lasioglossum fulvicorne	6	78
Lasioglossum laeve	**9**	98	Map 507
Lasioglossum laevigatum	4	102
Lasioglossum laticeps	5	90
Lasioglossum lativentre	**9**	100	Map 508
Lasioglossum leucopus	5	92
Lasioglossum leucozonium	7	90
Lasioglossum malachurum	5	94
Lasioglossum morio	5	96
Lasioglossum nitidiusculum	7	92
Lasioglossum parvulum	8	90
Lasioglossum pauxillum	5	98
Lasioglossum prasinum	4	104
Lasioglossum punctatissimum	8	92

Lasioglossum puncticolle	7	94	
Lasioglossum quadrinotatum	9	102	**Map 509**
Lasioglossum rufitarse	5	100	
Lasioglossum sexnotatum	6	80	
Lasioglossum smeathmanellum	5	102	
Lasioglossum villosulum	6	82	
Lasioglossum xanthopus	3	110	
Lasioglossum zonulum	6	84	
Lasius alienus	9	14	**Map 465**
Lasius brunneus	2	40	
Lasius flavus	8	16	
Lasius fuliginosus	1	22	
Lasius mixtus	9	16	**Map 466**
Lasius niger	9	18	**Map 467**
Lasius platythorax	9	20	**Map 468**
Lasius psammophilus	9	22	**Map 469**
Lasius sabularum	9	24	**Map 470**
Lasius umbratus	9	26	**Map 471**
Leptothorax acervorum	2	34	
Leptothorax albipennis	2	36	
Leptothorax interruptus	6	26	
Leptothorax nylanderi	6	28	
Leptothorax unifasciatus	6	30	
Leptothorax tuberum	2	36	
Lestica clypeata	9	38	**Map 477**
Lestiphorus bicinctus	3	84	
Lindenius albilabris	6	46	
Lindenius panzeri	6	48	
Macropis europaea	2	108	
Megachile centuncularis	8	102	
Megachile circumcincta	7	106	
Megachile dorsalis	7	108	
Megachile lapponica	9	118	**Map 517**
Megachile ligniseca	8	104	
Megachile maritima	3	114	
Megachile versicolor	8	106	
Megachile willughbiella	8	108	
Melecta albifrons	6	106	
Melecta luctuosa	6	108	
Melitta dimidiata	2	98	
Melitta haemorrhoidalis	2	100	
Melitta leporina	2	102	
Melitta tricincta	2	104	
Mellinus arvensis	2	94	

Mellinus crabroneus	2	96	
Methocha articulata	2	24	
Methocha ichneumonides	2	24	
Microdynerus exilis	3	52	
Mimesa bicolor	5	48	
Mimesa bruxellensis	5	50	
Mimesa equestris	5	52	
Mimesa lutaria	5	54	
Mimumesa atratina	5	56	
Mimumesa dahlbomi	5	58	
Mimumesa littoralis	5	60	
Mimumesa spooneri	5	62	
Mimumesa unicolor	5	64	
Miscophus ater	7	40	
Miscophus bicolor	7	42	
Miscophus concolor	7	44	
Monosapyga clavicornis	2	30	
Mutilla europaea	1	16	
Myrmecina graminicola	6	34	
Myrmica bessarabica	8	20	
Myrmica karavajevi	8	18	
Myrmica lobicornis	2	32	
Myrmica rubra	9	28	**Map 472**
Myrmica schencki	7	28	
Myrmica specioides	8	20	
Myrmica sulcinodis	3	24	
Myrmosa atra	2	26	
Nitela borealis	5	40	
Nitela lucens	5	42	
Nomada argentata	3	118	
Nomada armata	3	120	
Nomada baccata	9	82	**Map 499**
Nomada baeri	9	84	**Map 500**
Nomada castellana	9	84	**Map 500**
Nomada conjungens	7	110	
Nomada errans	4	108	
Nomada fabriciana	6	110	
Nomada ferruginata	4	110	
Nomada flavoguttata	9	86	**Map 501**
Nomada fucata	4	112	
Nomada fulvicornis	8	118	
Nomada goodeniana	9	88	**Map 502**
Nomada guttulata	5	110	
Nomada hirtipes	7	112	

Nomada integra	8	120	
Nomada lathburiana	4	114	
Nomada leucophthalma	3	122	
Nomada marshamella	8	122	
Nomada obtusifrons	5	112	
Nomada panzeri	8	124	
Nomada pleurosticta	8	120	
Nomada roberjeotiana	5	114	
Nomada ruficornis	**9**	90	**Map 503**
Nomada rufipes	6	112	
Nomada sexfasciata	6	114	
Nomada sheppardana	**9**	92	**Map 504**
Nomada signata	5	116	
Nomada striata	8	126	
Nomada xanthosticta	4	110	
Nysson dimidiatus	3	70	
Nysson interruptus	3	72	
Nysson spinosus	3	74	
Nysson trimaculatus	3	76	
Odynerus melanocephalus	1	40	
Odynerus reniformis	1	42	
Odynerus simillimus	1	44	
Odynerus spinipes	1	46	
Omalus aeneus	5	12	
Omalus puncticollis	5	14	
Osmia aurulenta	2	116	
Osmia bicolor	2	118	
Osmia caerulescens	5	106	
Osmia inermis	1	114	
Osmia leaiana	5	108	
Osmia parietina	4	106	
Osmia pilicornis	1	116	
Osmia rufa	2	120	
Osmia spinulosa	7	104	
Osmia uncinata	1	118	
Osmia xanthomelana	1	120	
Oxybelus argentatus	1	72	
Oxybelus mandibularis	1	74	
Oxybelus uniglumis	1	76	
Panurgus banksianus	7	82	
Panurgus calcaratus	7	84	
Passaloecus clypealis	4	60	
Passaloecus corniger	4	62	
Passaloecus eremita	4	64	

Passaloecus gracilis	4	66	
Passaloecus insignis	4	68	
Passaloecus monilicornis	4	70	
Passaloecus singularis	4	72	
Passaloecus turionum	4	74	
Pemphredon austriaca	8	48	
Pemphredon clypealis	8	54	
Pemphredon enslini	8	48	
Pemphredon inornatus	8	50	
Pemphredon lethifer	8	52	
Pemphredon lugubris	8	46	
Pemphredon morio	8	54	
Philanthus triangulum	1	98	
Philoctetes truncatus	5	16	
Plagiolepis taurica	6	24	
Plagiolepis vindobonensis	6	24	
Podalonia affinis	1	82	
Podalonia hirsuta	1	84	
Pompilus cinereus	3	34	
Ponera coarctata	7	32	
Priocnemis agilis	8	26	
Priocnemis confusor	9	30	**Map 473**
Priocnemis cordivalvata	8	28	
Priocnemis coriacea	4	32	
Priocnemis exaltata	8	30	
Priocnemis femoralis	8	32	
Priocnemis femoralis	9	32	**Map 474**
Priocnemis fennica	8	32	
Priocnemis gracilis	9	30	**Map 473**
Priocnemis hyalinata	9	32	**Map 474**
Priocnemis parvula	9	34	**Map 475**
Priocnemis perturbator	4	34	
Priocnemis pusilla	9	36	**Map 476**
Priocnemis schioedtei	4	36	
Priocnemis susterai	4	38	
Psen ater	5	66	
Psenulus concolor	7	60	
Psenulus pallipes	7	62	
Psenulus schencki	7	64	
Pseudepipona herrichii	3	48	
Pseudepipona tomentosus	3	46	
Pseudisobrachium subcyaneum	2	14	
Pseudocilissa dimidiata	2	98	
Pseudomalus auratus	5	18	

Name			
Pseudomalus violaceus	5	20	
Psithyrus species - see Bombus			
Pseudospinolia neglecta	5	22	
Rhopalum clavipes	6	50	
Rhopalum coarctatum	6	52	
Rhopalum gracile	6	54	
Rhopalum nigrinum	6	54	
Sapyga clavicornis	2	30	
Sapyga quinquepunctata	3	22	
Sifolinia karavajevi	8	18	
Smicromyrme rufipes	2	28	
Solenopsis fugax	7	30	
Sphecodes ephippius	**9**	104	**Map 510**
Sphecodes ferruginatus	**9**	106	**Map 511**
Sphecodes gibbus	5	104	
Sphecodes miniatus	**9**	108	**Map 512**
Sphecodes monilicornis	**9**	110	**Map 513**
Sphecodes niger	7	96	
Sphecodes pellucidus	8	94	
Sphecodes puncticeps	8	96	
Sphecodes reticulatus	6	86	
Sphecodes rubicundus	6	88	
Sphecodes ruficrus	6	88	
Sphecodes rufiventris	6	88	
Sphecodes scabricollis	6	92	
Sphecodes spinulosus	3	112	
Sphex funerarius	**9**	48	**Map 482**
Sphex rufocinctus	**9**	48	**Map 482**
Spilomena beata	**9**	40	**Map 478**
Spilomena curruca	**9**	42	**Map 479**
Spilomena enslini	**9**	44	**Map 480**
Spilomena troglodytes	**9**	46	**Map 481**
Stelis breviuscula	2	110	
Stelis ornatula	2	112	
Stelis phaeoptera	2	114	
Stelis punctulatissima	1	112	
Stenamma debile	8	22	
Stenamma westwoodii	8	24	
Stigmus pendulus	6	56	
Stigmus solskyi	6	58	
Symmorphus bifasciatus	1	48	
Symmorphus connexus	1	50	
Symmorphus crassicornis	1	52	
Symmorphus gracilis	1	54	

Tachysphex nitidus	2	70
Tachysphex obscuripennis	2	72
Tachysphex pompiliformis	2	74
Tachysphex unicolor	2	70
Tachysphex unicolor	2	76
Tapinoma erraticum	3	28
Temnothorax albipennis	2	36
Temnothorax interruptus	6	26
Temnothorax nylanderi	6	28
Temnothorax unifasciatus	6	30
Tetramorium caespitum	3	26
Tiphia femorata	3	18
Tiphia minuta	3	20
Trichrysis cyanea	6	22
Trypoxylon attenuatum	8	34
Trypoxylon clavicerum	8	36
Trypoxylon figulus	8	38
Trypoxylon medium	8	40
Trypoxylon minus	8	42
Vespa crabro	1	56
Vespula austriaca	2	64
Vespula germanica	4	48
Vespula rufa	3	58
Vespula vulgaris	4	50